Virtual You

Virtual You

How Building Your Digital Twin
Will Revolutionize Medicine
and Change Your Life

Peter Coveney
Roger Highfield

PRINCETON UNIVERSITY PRESS
PRINCETON AND OXFORD

Published by Princeton University Press
41 William Street, Princeton, New Jersey 08540
99 Banbury Road, Oxford OX2 6JX

press.princeton.edu

All Rights Reserved

Library of Congress Cataloging-in-Publication Data

Names: Coveney, Peter (Peter V.), author. | Highfield, Roger, author.
Title: Virtual you : how to build your digital twin in six steps—and revolutionize
    healthcare and medicine / Peter Coveney and Roger Highfield.
Description: First edition. | Princeton, New Jersey : Princeton University Press, [2023] |
    Includes bibliographical references and index.
Identifiers: LCCN 2022024293 (print) | LCCN 2022024294 (ebook) |
    ISBN 9780691223278 (hardback ; acid-free paper) |
    ISBN 9780691223407 (e-book)
Subjects: LCSH: Virtual reality in medicine. | Virtual humans (Artificial intelligence) |
    Medicine—Computer simulation. | Computer vision in medicine. |
    BISAC: SCIENCE / Chemistry / Physical & Theoretical | COMPUTERS /
    Computer Science
Classification: LCC R859.7.C67 C69 2023 (print) | LCC R859.7.C67 (ebook) |
    DDC 610.285—dc23/eng/20221109
LC record available at https://lccn.loc.gov/2022024293
LC ebook record available at https://lccn.loc.gov/2022024294

British Library Cataloging-in-Publication Data is available

Editorial: Ingrid Gnerlich and Whitney Rauenhorst
Production Editorial: Natalie Baan
Text and Jacket Design: Chris Ferrante
Production: Jacquie Poirier
Publicity: Kate Farquhar-Thomson, Sara Henning-Stout, and Maria Whelan
Copyeditor: Jennifer McClain

This book has been composed in IBM Plex

Printed on acid-free paper. ∞

Printed in the United States of America

10 9 8 7 6 5 4 3 2 1

For all those doctors, scientists and engineers
who are making medicine truly predictive

Contents

Foreword

Venki Ramakrishnan

The physicist Richard Feynman once said, "What I cannot create, I do not understand." One could add, "What I cannot accurately simulate, I do not understand." We can be confident we understand how airplanes fly and how they behave during engine failure or turbulence because a flight simulator can accurately predict what would happen. Indeed, pilots are routinely trained both for normal and unexpected situations using flight simulators, providing them with the experience that they might never encounter using real aircraft.

What are the prospects of simulating ourselves in a computer? At first sight, this sounds more science fiction than fact. However, much progress has been made; and this book takes us on a grand tour of the first steps towards the creation of our own digital twin and the challenges in accomplishing this goal, which is a fascinating journey in itself.

One of the key themes of the book is an aspiration to see as much theory used in biology as in physics. When I left physics in 1976 immediately after graduate school to retrain all over again as a biologist, I immediately noticed a difference. Physics as well as chemistry had highly advanced theory that helped guide and even predict experiments and behaviour. So, from theoretical considerations, it was possible to invent transistors and lasers and synthesize entirely new compounds. At the extreme, it even led to the construction of a multibillion-dollar accelerator to look for the theoretically predicted Higgs boson. It is far from perfect. For example, there is no good theory for high-temperature superconductivity, and we cannot predict the detailed superconducting behaviour of a mixture of metals.

By contrast, certainly in the 1970s, biology seemed largely observational and empirical. It is true that it did have one encompassing theory—that of natural selection acting as a driving force for the evolution of all life. While this theory had great explanatory power, it did not have detailed predictive power. Biology also had what I would call local theories or models: an understanding of how an impulse travels along a nerve, how various biological motors work, or how you would respond to a shot of adrenaline. But what it lacked was an overarching ability to predict from a set of initial conditions how a system, even a basic unit like a cell, would behave over time under an arbitrary set of conditions.

Part of the problem is that the number of factors involved in sustaining life is enormous. At some basic level, there is our genome, whose sequence was announced at the beginning of this century with much fanfare and the promise of ushering in a new age of biology. The genome consists of tens of thousands of genes, all expressed in varying degrees and at varying times. Moreover, the expression of genes is modulated by chemical tags added to the DNA, or proteins associated with it, that persist through cell division—the subject of epigenetics. Finally, the expression of genes and their function in the cell is the result of their interaction with each other and the environment to form a giant network. So, any early notions of being able to predict even just gene-based outcomes by knowing someone's sequence turned out to be far-fetched. If this is the case for just a single cell, one can imagine the levels of complexity in trying to predict outcomes theoretically for an organ, let alone an entire human being.

Over the last 20 years, the biological revolution of the twentieth century has continued, but this time augmented by a revolution in computing and data. We now have vast amounts of data. Instead of a single human genome sequence, we have hundreds of thousands, as well as vast genome sequences of an enormous number of species. We have transcriptome maps that tell us which genes are expressed (the process by which the information encoded in a gene is put to work) in which cells and when, and we have interactome maps that map the interaction between the thousands of genes. Along with this we have large-scale data on human physiology and disease. We have data on the personal health characteristics of millions of individuals. All of these data are simply beyond the ability of any person or even a group to analyse and interpret. But at the same time, data science has also exploded. Computational techniques to link large data sets and make sense of them are advancing all the time.

Will these advances, in conjunction with a robust development of theory, eventually allow us to simulate a life-form? This book argues that the confluence of these developments is allowing us to tackle the ambitious goal of simulating virtually every organ and every process in a human. When that is done, one can begin to ask questions: How will this person respond to a particular treatment? At what point is someone likely to develop a debilitating illness? Could they prevent it by taking action early on and, if so, what action is likely to work? How would one person's immune system respond to an infection compared to someone else's?

The book describes progress in virtually every area of computational biomedicine, from molecules and cells to organs and the individual. Reading it, I get the impression that there are areas where a digital simulation—a virtual copy of the real thing—is almost here. Other goals seem plausible, and there are yet others that seem a lot like science fiction. Yes, in principle it might be possible to achieve them eventually, but the practical difficulties seem enormous and currently insurmountable. Scientists may well debate which parts are a fantasy and which parts are a glimpse into our future. Coveney and Highfield themselves draw the line at digital consciousness.

Those of us on the more sceptical end should remember that there are two almost universal characteristics of new technologies. The first is that, in the initial stages, they do not seem to work well and their applications seem very narrow and limited. In what seems to be an abrupt transformation, but is really the result of several decades of work, they suddenly become ubiquitous and are taken for granted. We saw that with the internet, which for the first couple of decades was the preserve of a few scientists in academia and government labs before it exploded and dominated our lives. A second characteristic, pointed out by Roy Amara, is that when it comes to predicting the effect and power of a new technology, we tend to overestimate the short term and underestimate the long term. This often leads to great initial hype followed by the inevitable disappointment followed by eventual success. For example, the hype around artificial intelligence in the 1970s and '80s proved quite disappointing for decades, but today the field is making enormous strides. Perhaps we are seeing the same trajectory with driverless cars.

Whether the same will be true for the creation of virtual humans remains a matter of debate, but in this book Coveney and Highfield offer us a fascinating account of the diverse efforts around the world that are now under way with that extraordinary goal in mind.

Venki Ramakrishnan shared the Nobel Prize in 2009 for his studies of the ribosome, the molecular machine that turns genes into flesh and blood.

Virtual You

Introduction

Imagine a virtual human, not made of flesh and bone but one made of bits and bytes, and not just any human, but a virtual version of you, accurate at every scale, from the way your heart beats down to the letters of your DNA code.

—*Virtual Humans* movie premiere, Science Museum, London

Within the walls of a nineteenth-century chapel on the outskirts of Barcelona, a heart starts to contract. This is not a real heart but a virtual copy of one that still pounds inside a patient's chest. With its billions of equations, and 100 million patches of simulated cells, the digital twin pumps at a leisurely rate of around one beat per hour as it tests treatments, from drugs to implants.

Though it was deconsecrated many decades ago, the Chapel Torre Girona is still adorned with a cross above its entrance. You can sense a higher power and purpose inside its romantic architecture. There, as sunlight streams through its stained-glass windows, you are confronted by an enormous glass-and-steel room, within which stand three ranks of black cabinets dotted with green lights.

This is MareNostrum (the Roman name for the Mediterranean Sea), a supercomputer on the campus of the Polytechnic University of Catalonia that is used by Peter Coveney along with colleagues across Europe to simulate electrical, chemical, and mechanical processes within the human body. These simulations look just like the real thing, whether a fluttering heart or a lung expanding into the chest. Much more important, however, is that these virtual organs behave like the real thing.

To show the dazzling range and potential of virtual human research, we used MareNostrum to create a movie, with the help of simulations run on other supercomputers, notably SuperMUC-NG in Germany (the suffix *MUC* refers to the code of nearby Munich Airport). Working with an international team, we wanted our *Virtual Humans* movie to showcase where these diverse efforts to create a body *in silico* could take medicine.

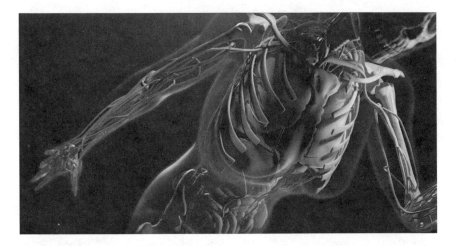

FIGURE 1. Still from the *Virtual Humans* movie. (CompBioMed and Barcelona Supercomputing Centre)

In September 2017, we held the premiere in the cavernous IMAX Cinema of the Science Museum in London with Fernando Cucchietti and Guillermo Marin, our colleagues from the Barcelona Supercomputing Centre. Even though we had worked for many months on the movie, gazing up at a pounding virtual heart the size of four double-decker buses still left us a little breathless.

SuperMUC-NG and MareNostrum 4 are among a few hundred or so great computational machines dotted around the world that are being harnessed to model the cosmos, understand the patterns of nature and meet the major challenges facing our society, such as studying how the Earth will cope with climate change, developing low-carbon energy sources and modelling the spread of virtual pandemics.

Just as great medieval cathedrals were raised by architects, masons, geometers and bishops to give humankind a glimpse of the infinite, supercomputers are the cathedrals of the information age, where novel worlds of endless variety, even entire universes, can be simulated within these great engines of logic, algorithms and information.

You can also re-create the inner worlds of the human body, and not just any body, or an average body, but a particular person, from their tissues and organs down to the molecular machines at work within their cells, their component proteins along with their DNA.

FIGURE 2. The MareNostrum supercomputer. (Wikimedia Commons: Gemmaribasmaspoch. CC-BY-SA-4.0)

The eventual aim of this endeavour is to capture life's rhythms, patterns and disorders in a computer, not just of any life or an average life, but of one particular body and one particular life—yours.[1]

At the premiere, we were joined by colleagues who had developed virtual hearts, arteries and veins along with the skeleton and its musculature. On that great IMAX screen in the Science Museum, the packed audience glimpsed a future when drugs can be designed to suit an individual patient, when we can visualise the shimmering movements of a mutated protein in the body, track the turbulent flow of drug particles deep into the lungs, study the surges of blood cells through the brain, and simulate the stresses and strains that play on weakened bones.

Rise of Digital Twins

In engineering, virtual copies are known as digital twins. The concept is usually attributed to a paper by John Vickers and Michael Grieves at the University of Michigan in 2002,[2] which talked of a "Mirrored Spaces Model." NASA coined the term *digital twin* in 2010,[3] and applied this way of thinking to spacecraft.[4] However, the origins of this approach can be glimpsed much earlier. Many cite the Apollo moon programme as one notable example, when simulators on the ground were used as analogue twins of spacecraft. This approach was famously employed in 1970 to help return three astronauts safely to Earth in the aftermath of an explosion 200,000 miles out in space on board the ill-fated Apollo 13 mission.[5]

Today, digital twins are well established. Many industrial processes and machines are too complex for one brain to grasp, so experimenting with their digital twins makes their behaviour easier to explore and understand.[6] Lessons learned this way are transforming the future of manufacturing and, by accelerating automation, altering the future of work. Digital copies of machines, even entire factories, are helping to anticipate hurdles, perfect designs and prevent mistakes before they occur.

Digital twins are used to optimise supply chains and store layouts; General Electric used a twin to boost efficiency at an aluminium smelter in India; a twin of the route of a proposed railway line in north west England—in the form of 18 billion data points harvested by drones—was created to help manage this vast transport project; a "factory of the future" in Australia honed a virtual copy of a robotic workstation before building the real thing; engineers use digital twins to estimate the lifetime of a jet engine and how to maintain it efficiently. Digital twins have been used to help create wind turbines, oil rigs, cars, jet engines, aircraft, spacecraft and more besides. Some believe that digital twin cities hold the key to future urban planning.

Digital twins are emerging in medicine too, thanks to the data revolution in biology. One of the legions of people sifting through health data is Leroy Hood of the Institute for Systems Biology, Seattle. Among the most influential of today's biotechnologists, Hood has worked at the leading edge of medicine, engineering and genetics for decades, dating back to the first human genome programme meeting in 1985. In 2015, he launched a venture that gathered a plethora of data on 5000 patients for five years. All their data were stored in what Hood calls "personal health clouds."* Analysis of a patient's cloud can reveal telltale signals of what Hood calls "pre-pre-disease" that doctors could use to anticipate problems, then intervene to maintain their health.

Hood talks of "scientific wellness," which "leverages personal, dense, dynamic data clouds to quantify and define wellness and identify deviations from well states toward disease." A living embodiment of his approach, the 82-year-old was on sprightly form ("I plan never to retire") when we talked to him about his vision of a "P4" future, where treatments are predictive, preventive, personalised, and participatory. Simulations of the body will help usher in that future by making sense of what patterns in a patient's data hold in store for them.

In reality, of course, we make do with incomplete understanding and incomplete data. But, as advances in weather forecasting have shown, these shortcomings can be overcome to make useful predictions. We have come a long way since 1922 when, in his remarkable book *Weather Prediction by Numerical Process*, the British mathematician Lewis Fry Richardson (1881–1953) outlined the idea of a fantastic forecast factory, where thousands of human "computers," using slide rules and calculators, are coordinated by a "conductor." Richardson mused on whether "some day in the dim future it will be possible to advance the computations faster than the weather advances." But even he went on to admit that his forecast factory was only a dream.

A century later, his extraordinary vision has become a reality. Supercomputers can make predictions a few days into the future with reasonable accuracy by constantly updating sophisticated computer models with data from orbiting satellites, buoys, aircraft, ships and weather stations.

A typical forecasting model relies on a system of equations to simulate whether it is going to rain or shine. There is an equation for momentum, density, and temperature in each of water's three phases (vapour, liquid and solid), and potentially for other chemical variables too, such as the ozone that absorbs harmful ultraviolet radiation. In

* Leroy Hood, interview with Peter Coveney and Roger Highfield, August 12, 2021.

Chapter Two, we spell out why these nonlinear differential equations, notably partial differential equations, rule the climate system. In all, it takes *billions* of equations to model the planet down to a resolution of, currently, around 60 kilometres.* Overall, the model has to take account of ever-changing thermodynamic, radiative and chemical processes working on scales from hundreds of metres to thousands of kilometres, and from seconds to weeks.[7] That represents a tour de force of simulation, one that some claim already approaches the complexity required to model the human brain.

Thanks to the torrent of biomedical data available today, along with ever more powerful theory and computation, we believe simulations will revolutionise biology just as much as they have transformed meteorology. The American meteorologist Cleveland Abbe (1838–1916) once declared how progress in his field depended on "the consecration of the physicist and mathematician to this science."[8] To echo his 1895 vision of forecasting, we look forward to the day when it is not enough to know someone is unwell—we want to be able to understand if they will fall sick and why, so that we can make them better and for longer.

Optimism about the potential of digital twins in medicine is bolstered by our current ability to forecast weather, which would amaze Abbe. We take the daily forecasts for granted, but this feat of prediction is truly extraordinary. Markus Covert of Stanford University, who has developed virtual cells, remarked that "prediction of storms such as Hurricane Sandy ten days in advance of landfall—with the corresponding evacuation of hundreds of residents, saving both lives and property—could arguably be ranked as among the great technical triumphs in human history."[9]

When it comes to climate forecasts, plans are under way to create a "digital twin" of Earth that would simulate the atmosphere, ocean, ice, and land down to a resolution of one kilometre, providing forecasts of the risks of floods, droughts, and fires, along with the swirling ocean eddies that shift heat and carbon around the planet. This European model, Destination Earth, will fold in other data, such as energy use, traffic patterns and human movements (traced by mobile

---

* Tim Palmer, email to Peter Coveney, June 2, 2021.

phones), to reveal how climate change will affect society—and how society could alter the trajectory of climate change in what some already call the Anthropocene, a geological epoch where human activity is having a significant impact on our planet.[10]

The details of creating a digital twin of our own planet Earth are staggering. Take clouds, as one example. They are made of water, which is also the main ingredient of the human body (around 68%[11]). Unlike us, however, clouds seem simple—great plumes of water droplets or ice crystals floating in the sky. Their formation is critical to our ability to predict weather, important for our understanding of the effects of global heating and central to controversial schemes to curb climate change through geoengineering.[12]

From cumulus tufts with beguiling shapes to great sheets of grey, clouds are a beautiful example of how complexity can result from simplicity, as droplets of water are borne on air currents of convection. As these droplets condense inside clouds, a little heat is released, making the clouds buoyant. At great heights, where temperatures fall well below freezing, the droplets turn into ice crystals, giving the resulting cirrus clouds a wispy, feathery look.

Within a cloud, processes at the smallest scales govern the formation of droplets. But, though microscopic, these features and interactions have large-scale, macroscopic, effects. The smaller and more numerous the droplets, the more that light is scattered. At the scale of micrometres, turbulence accelerates cloud formation and triggers rain showers.[13] Large-scale air motions can create vast cloud systems that can span a continent. By reflecting light into space, clouds can cool the Earth's surface, which is why some believe they should be nurtured to help curb runaway global warming.[14]

Essentially all the laws that underpin cloud formation are known, so we should be able to represent how they evolve in terms of known mathematical equations. The hope is to achieve the same for virtual humans, even down to the last water molecule. This may sound fantastical, but optimism that mathematics can describe the warm, complex, dynamic world of the body dates back centuries. The English physician William Harvey (1578–1657) relied on calculations in his demonstration of the circulation of the blood,[15] while in 1865 the French physiologist Claude Bernard (1813–1878) stated that "the application of mathematics to natural phenomena is the aim of all science."[16]

Our ability to create a virtual copy of a person depends on describing the body with the language of mathematics. Although a work in progress, equations written using calculus, which express rates of change, can already depict complex processes uncovered by molecular biologists, cell biologists and many others in the biosciences. These mathematical expressions—ordinary and partial differential equations—can describe at every instant how blood pressure varies depending on where you make a measurement in the body or track an electrical impulse as it speeds along a neuron in the brain, or how quickly a virus steals into a person's airway.

To put these equations to work, all that is needed to start calculating are the boundary conditions for the problem at hand. This could mean the state of a neuron or an infected cell at a given time or at various time intervals, their rates of change at various instants or the upper and lower limits of a given quantity. These conditions tether the mathematics to reality so we can make forecasts about the body, or "healthcasts," by analogy with the weather.

But while we accept that the laws of nature are universal, in one critical and practical sense the life sciences—by which we mean biology and medicine—are quite different from the physical sciences—physics and chemistry—that we use to describe clouds. They are more empirical, more dependent on making measurements and doing experiments and, until now, less dependent on theoretical understanding.

Theory, that is, the mathematical representation of the laws of nature, plays a relatively diminished role in medicine and biology. Even the Darwin-Wallace theory of evolution, regarded by some as the greatest scientific theory of all, does not admit a mathematical description. This might sound shocking, but the reality is that, while basic predictions about the patterns of inheritance have been made since Gregor Mendel studied peas in the nineteenth century, the course of evolution is not possible to predict in any quantitative manner.[17]

Some influential figures are only too aware of this shortcoming. Paul Nurse, director of the Francis Crick Institute in London and former assistant editor of the *Journal of Theoretical Biology*, told us how he was weary of reading papers that use clever technology to make measurements that come to "barely any significant conclusions."* In an opinion article for the journal *Nature*, he cited Sydney Brenner (1927–2019), his old friend and fellow Nobelist: "We are drowning in a sea of data and starving for knowledge."[18] He complained to us that the importance of theory and the principles of life are relatively neglected in favour of cramming facts, knowledge and information. Biology "does have ideas, so why aren't we talking about them?"

Yet biology, like the rest of science, is undoubtedly governed by the laws of nature. To be sure, there are no-go areas for moral and ethical reasons based on human arguments, but there is absolutely every reason to believe that we should be able to understand a particular scientific aspect of how an organism works and capture that insight in the form of mathematics. To create Virtual You, we need to go beyond the current use of theory in making post hoc rationalisations in biology, after studies are carried out, to using theory to guide experiments and make predictions.

Uniting Science

Science is balkanised. The notion of dividing academic inquirers into tribes dates back to ancient Greece with Socrates (c. 469–399 BCE), his student Plato (c. 428–347 BCE) and, in turn, Plato's student Aristotle (384–322 BCE).[19] Within a few decades, however, Timon of Philius (c. 320–230 BCE) moaned about the squabbling of "bookish cloisterlings" at the Museum of Alexandria. By the sixteenth century, Francis Bacon (1561–1626) and other philosophers were mourning the splintering of human knowledge.

By the mid-nineteenth century, the disciplinary boundaries of the modern university had taken root, each with its own customs, language, funding streams, establishments and practices. In *Virtual You*, we intend to show that today's research is more than a baggy collection of fragmented efforts—it is a grand and complementary mosaic of data, models, mechanisms and technology. The big picture of how the human body works is beginning to heave into view.

* Paul Nurse, interview with Peter Coveney and Roger Highfield, September 25, 2021.

Just as there is no privileged point of view of the human body, so each perspective from each discipline is equally important. Each is complementary and, if united and consistent, remarkable new insights can emerge. If we look, for example, at the great molecular biology revolution that dates from the 1950s, when physicists and chemists tackled biology, and biologists used techniques developed by physicists, we can see that this vital atomic view of proteins, enzymes and other molecules of living things perfectly complements existing insights into heredity and evolution, marking a powerful unification of knowledge known as consilience.

The simple idea at the heart of this book is that the convergence of many branches of science—patient data, theory, algorithms, AI and powerful computers—is taking medicine in a new direction, one that is quantitative and predictive. We will show how mathematics can capture an extraordinary range of processes at work in living things, weigh up developments in computer hardware and software and then show how the human body can be portrayed *in silico*, holding up a digital mirror to reflect our possible futures.

This is a story that builds on multidisciplinary ideas we set out in our earlier books, *The Arrow of Time*[20] and *Frontiers of Complexity*.[21] In the first, we discussed how to reconcile a deep problem at the heart of science: that time is represented in different ways by different theories and at different length scales, ranging from the microscopic to the macroscopic. In the latter, we showed how complexity in mathematics, physics, biology, chemistry and even the social sciences is transforming not only the way we think about the universe, but also the very assumptions that underlie conventional science, and how computers are essential if we are to explore and understand this complexity. Nowhere is this more relevant than in the efforts to create the virtual human. In *Virtual You*, we draw these threads together within a broad tapestry of research, both historical and contemporary.

Virtual You

This is the first account of the global enterprise to create a virtual human aimed at the general reader. Hundreds of millions of dollars have been spent in the past two decades on the effort that has been organised through initiatives such as the International Physiome Project,[22] America's Cancer Patient Digital Twin,[23] the European Virtual Physiological Human,[24] the Human Brain Project[25] and another Europe-wide effort led by University College London to which we both contribute, Computational Biomedicine, or CompBioMed for short.

All are united by a single objective. As one workshop held in Tokyo declared: "The time is now ripe to initiate a grand challenge project to create over the next 30 years a comprehensive, molecules-based, multi-scale, computational model of the human ('the virtual human'), capable of simulating and predicting, with a reasonable degree of accuracy, the consequences of most of the perturbations that are relevant to healthcare."[26] That virtual vision was unveiled more than a decade ago—in February 2008—and its future is fast approaching.

In the following pages, we will take you on a fantastic voyage through the body, its organ systems, cells and tissues along with the deformable protein machines that run them. We hope to convince you that, in coming decades, virtual twins of cells, organs, and populations of virtual humans will increasingly shape healthcare. This organising principle for twenty-first-century medicine will enable doctors for the first time to look forward to—and predict—what is in store for you, including the effects of proposed therapies. This marks a stark contrast with today's approach where doctors, in effect, look back at what happened to similar (though nonidentical) patients in similar (though nonidentical) circumstances.

In the long term, virtual cells, organs and humans—along with populations of virtual humans—will help to evolve the current generation of one-size-fits-all medicine into truly personalised medicine. Your digital twin will help you understand what forms of diet, exercise and lifestyle will offer you the healthiest future. Ultimately, the rise of these digital twins could pave the way for methods to enhance your body and your future. As we discuss in our concluding chapter, virtual humans will hold up a mirror to reflect on the very best that you can be.

The following four chapters focus on the fundamental steps that are required to create a digital twin: harvest diverse data about the body (Chapter One); craft theory to make sense of all these data

(Chapter Two) and use mathematics to understand the fundamental limits of simulations; harness computers to put the spark of life into mathematical understanding of the human body (Chapter Three); blend the insights of natural and artificial intelligence to interpret data and to shape our understanding (Chapter Four).

In Chapters Five to Eight, we show the consequences of taking these steps and begin to build a digital twin, from virtual infections (Chapter Five) to cells, organs, metabolism and bodies. Along the way, in Chapter Six, we encounter the fifth step necessary for the creation of Virtual You. Can we stitch together different mathematical models of different physical processes that operate across different domains of space and time within the body? We can, and the ability to customise a virtual heart to match that of a patient marks one extraordinary example (Chapter Seven), along with modelling the body and its organ systems (Chapter Eight). In Chapter Nine, we discuss "Virtual You 2.0," when the next generation of computers will overcome shortcomings of the current generation of "classical" digital computers.

In our last chapter, we examine the many opportunities, along with ethical and moral issues, that virtual humans will present. Digital twins will challenge what we mean by simple terms such as "healthy." Are you really healthy if your digital twin predicts that—without a treatment or a change in lifestyle—you will not live out your potential life span? You may feel "well," but are you really well if simulations suggest that you are destined to spend a decade longer in a care home than necessary? If a virtual human can become the substrate for human thought, how will we come to regard our digital copy? Finally, in an appendix, we examine a provocative question raised by using computers to simulate the world: Is it possible to re-create the fundamental physics of the cosmos from simple algorithms?

So, to the first of our foundational chapters. This poses the most basic question of all. If we are to create digital twins, how well do we have to know ourselves? To create Virtual You, we need to understand how much data and what kinds are sufficient for a digital twin to be animated by a computer.

As Aristotle once remarked, knowing yourself is the beginning of all wisdom.

FIGURE 3. Virtual anatomical twin. One of the detailed high-resolution anatomical models created from magnetic resonance image data of volunteers. (IT'IS Foundation)

1

The Measure of You

In that Empire, the Art of Cartography attained such Perfection that the map of a single Province occupied the entirety of a City, and the map of the Empire, the entirety of a Province. In time, those Unconscionable Maps no longer satisfied, and the Cartographers' Guilds struck a Map of the Empire whose size was that of the Empire, and which coincided point for point with it.

—Jorge Luis Borges, "Del rigor en la ciencia"
(On Exactitude in Science)[1]

To create a virtual version of your body, the first step is to gather enough personal data. There are plenty of potential sources: ultrasound scans of your heart and other internal organs, or whole-body imaging, using X-rays or magnetic resonance imaging (MRI). You can draw on various -omes, whether your detailed DNA sequence (your genome) or the chemical details of your metabolism (your metabolome) or your entire complement of proteins (your proteome). Your personal data might include unfamiliar features, such as the particular shape of an important enzyme in your body, more routine measurements, such as blood pressure, along with "digital biomarkers" that can be harvested with a wearable device, be it a phone, a watch or a shirt made of smart textiles that monitor perspiration.[2]

But how much and what kind of data do we need, exactly? One answer comes from the single-paragraph story quoted above by the Argentine essayist Jorge Luis Borges. In this briefest of brief tales, Borges conjures up a time where the science of cartography becomes so precise that only a map on the same scale as an empire itself would suffice.

To what extent does science need to represent the human body in order to understand it? When it comes to creating a virtual you, do

we, like Suárez Miranda, have to capture the details of all the 7,000,
000,000,000,000,000,000,000,000 (7 octillion) atoms in your body,
let alone all the details of the even greater congregation of simpler
particles—spinning protons, neutrons, and electrons—that constitute
each of your component atoms? When considering how much data
we need to take the first step to create a digital twin, can we avoid
drowning in data, the curse of the cartographers' guild?

There are other issues to consider. We seek data that can be measured by anyone anywhere, using the same equipment, in the same
conditions and obeying the same protocols. Even different people
using different equipment should come up with similar findings in
similar conditions.[3] We need to gather these data in an efficient and
timely way: science is always spurred on by the development of new
instruments, such as microscopes, sequencers and scanners. There
are also issues about curating, storing and protecting your data. And,
of course, there are practical questions about processing all these
data: even the most powerful computer envisaged in the coming decades will be unable to simulate the behaviour at the molecular level
of the human body, which estimates suggest consists of somewhere
between 20,000,000,000,000,000,000,000,000,000 and 1,000,000,000,
000,000,000,000,000,000 molecules.

The State of You

Intuitively, it seems reasonable to assume that we need to know
everything we can about you if we are to create a virtual version.
But it is a tall order to measure the state of all your molecular ingredients, let alone all your component atoms. Equally, how much data
is enough? Would it be enough to know that your body is built with
around 20,000 genes? Or that it is a remarkable collective of 37.2 trillion cells?[4] Or that your brain weighs 3 pounds and requires around
20 watts of energy? Or that the molecules in your body are various
concoctions of some 60 or so atoms of different kinds (elements),
including 25 grams of magnesium, found in bones and muscle, 1.6
milligrams of cobalt, found in vitamin B_{12}, 4 milligrams of selenium
and 96 grams of chlorine?[5] Or that you need about 10 to the power of

11 bits (100,000,000,000 bits) to describe a scan of your body down to a length of 1 millimetre? Or that you would need 10 to the power of 32 bits (a one followed by 32 zeros) of information to describe your body at atomic resolution?

Data are not equal. Particularly telling data include "emergent" properties that reflect the collective behaviour of a large number of microscopic constituents, where the sum is qualitatively different from the behaviours of the parts. We began our first book with the Austrian physicist Ludwig Boltzmann (1844–1906), who showed how properties of fluids and gases emerged from the behaviour of their component molecules, helping to usher in a field that today is called statistical mechanics. Peter Sloot, who works with Peter at the University of Amsterdam, describes emergence in terms of interacting elements that adapt to an environment that they themselves help create.* Paul Nurse sums up emergence in a complementary way: higher levels of biological description, say, at the level of the cell, constrain events that take place at lower levels, such as among the molecules of life. "As a consequence," he told us, "you could never build life simply from the bottom up."†

Of the many examples of emergence in biology, where the whole is greater than the sum of the parts, the most vivid are life itself and consciousness. While a brain can be happy, its component neurons are unfettered by emotions. Similarly, while a bacterium is alive, its component molecules are not. Even if we knew the full molecular details of an organism, down to the last atom, we would not be able to tell that this is the recipe of a living thing.

The corollary of emergence is that it is not practical, necessary or even sufficient to carry knowledge of everything from one level of description (the octillions of atoms comprising the body) to another level, such as a single cell. And even if we did try to model the movements of the heart by starting at the atomic level, we would discover that it could take an eternity to simulate, even with the most powerful of computers. There is no point in creating a perfect—in the

* Peter Sloot, email to Peter Coveney and Roger Highfield, August 4, 2021.
† Paul Nurse, interview with Peter Coveney and Roger Highfield, September 24, 2021.

reductionist sense—cardiac model down to the last atom if a single virtual heartbeat of this model takes millennia.

The science of complexity also tells us that we don't need to know every last detail.[6] We intuitively know this because medicine may sometimes home in on elemental matters—such as sodium or iron levels—but diagnosis usually focuses on higher levels of description, from X-rays of bones to blood pressure and heart rate. To understand the science of the human, the amount of data that we need from lower—that is, smaller—levels is far less than you would naively expect. Moreover, by focusing on every last twig, branch and tree, we can easily miss the forest.

Though our knowledge of how the human body works is dependent on understanding its component parts, it is critical to grasp how all these pieces work together if we are to fathom its emergent properties. Even if we understand DNA's role in a cell (at present we only understand a tiny fraction of its function), and that cell's function in an organ, that does not mean we can figure out an organism's physiology because each cell is affected by the activity of cells in other tissues, organs and organ systems. When it comes to pathogens, such as viruses, we also need to understand how they can move from organism to organism, as in a pandemic. And then there are the ways in which organisms interact with one another, whether a virus in a host, or a person in a village, or a constituency in a society, a huge subject in its own right. Roger Highfield has coauthored an entire book—*Supercooperators*—about how and why humans are the most cooperative species of all.[7]

But even beyond that, these layers of organisation are affected by the environment, diet and lifestyle, such as exposure to sunlight, stress, fast food and exercise. From the earliest days of modelling human physiology, we have found evidence of "downward causation," that is, of how these and other influences on higher levels of organisation of the body can alter the way that genes are used in cells. We may carry genes that increase our risk of developing type 2 diabetes, but if we eat a healthy diet and get enough exercise, we may not develop the disease. Similarly, someone may carry genes that reduce his or her risk of developing lung cancer, but chain smoking may still be calamitous. Human biology is more than the sum of nature and nurture.

From Order to Chaos

Emergence of novel, organised attributes and structures from an interacting system of cells, tissues and organs in a given environment is the light side of complexity theory. However, there is a dark side of complexity in the form of what is called deterministic chaos. This puts another constraint on the extent to which we can turn data about the human body into understanding.

Deterministic chaos is not the same as randomness. In reality, it is a subtle form of order, one freed from the shackles of periodicity and predictability. Chaos can emerge from deceptively simple-looking equations that contain a key ingredient, nonlinearity, where a change in output is not proportional to a change in input. Examples of nonlinearity abound, from the nudge in temperature that triggers a boiler to cut out to the howlround caused when a microphone is brought a little too close to a speaker. Nonlinearity can lead to chaos, where predicting exact behaviour is impossible over the longer term.

Chaos is commonplace, from the unpredictable swing of a pendulum to the vagaries of the weather.[8] Chaos lurks within the body too. The problem with deterministic chaos is that, unless you feed in data with infinite precision (which is impossible), these complex nonlinear interactions make exact long-term predictions impossible. So, while you do not need to know everything about the body to model it, small changes in some data can lead to big, unpredictable outcomes, when we have to couch our predictions in terms of probabilities.

Instrumental Data

To work out the extent to which we want to capture this complexity of the human body and take that first step towards Virtual You, we need to think of data as an *instrument*, not as a representation. For the same reason, maps vary in detail depending on what you do with them: a hiker needs to see every field and footpath, while the pilot of a plane needs a chart of terrain, airports, air spaces and beacons. By the same token, the level of detail we require to take the first step to create Virtual You depends on what questions we want to ask.

A simple measurement, such as taking a temperature, may be all we need to figure out if a child has picked up an infection. But in elderly people, we might need more detailed data about how they are responding to an infection to understand what is going on. When it comes to a serious urinary tract infection, for example, the first sign of a problem could be confusion rather than a fever. And if we need to understand what kind of infection is causing the problem, then we do need more and different data, such as the genetic makeup of the infectious organism.

Data are truly instrumental when they are grounded in the scientific method, the most powerful way to provide rational insights into the way the body works. This relies on theory; otherwise science would be no more than the cataloguing of reproducible observations. As we mentioned in the introduction, equations in a typical theory represent nature's workings in a more economical manner than great repositories of raw data. Theory helps us uncover principles along with laws that explain how and why the body works as it does. We have more to say about this in the next chapter, when we discuss the second step towards Virtual You. First of all, however, we have to find ways to harvest data from the body.

A Brief History of Anatomy

"So much progress depends on the interplay of techniques, discoveries and new ideas, probably in that order of decreasing importance." When Nobelist Sydney Brenner made this remark in March 1980 at a symposium organised by the Friedrich Miescher Institute in Basel, Switzerland, he was looking forward to the next decade of biology.[9] As Brenner predicted, the development of the virtual human has been driven by new techniques, and its rise will continue to rely on more, new and different kinds of data. When it comes to the structures of the body, an array of techniques has revealed unprecedented details.

There are too many to mention them all, though many remarkable ways to view the body have emerged down the centuries. One has to be the publication in 1543 of *De Humani Corporis Fabrica* (On the Fabric of the Human Body), an extraordinary 700-page work

FIGURE 4. Portrait of Vesalius from his *De Humani Corporis Fabrica* (1543). (Created by Jan van Calcar)

by Andreas Vesalius (1514–1564), which presented more than 200 woodcuts based on his dissections of human bodies.

To amplify the insights of dissection and traditional anatomy, a vast array of techniques has been developed. *Micrographia*, the first important work on microscopy, was published in 1665. In this pioneering scientific best seller, Robert Hooke (1635–1703) revealed

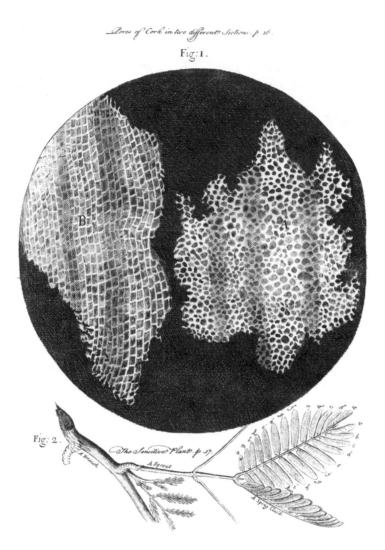

FIGURE 5. Cell structure of cork created by Robert Hooke, *Micrographia*, 1665.

the microscopic structure of cork. He showed walls surrounding empty spaces and referred to these structures as cells. Today, microscopy technologies can resolve the details of cells down to the atomic scale.[10]

We no longer have to study the body with only visible light. In 1895, the German physicist Wilhelm Röntgen (1845–1923) discovered an invisible kind of ray, called X-radiation or X-rays. That

Christmas, he wrote a 10-page article in which he described how X-rays could make bones visible. These revealing rays would also make it possible to study the molecular machinery of cells, through a technique called X-ray diffraction. Today, an array of other methods can peer inside the living body, from terahertz imaging to ultrasound. We can even use antimatter, in the form of positrons (antielectrons), to study metabolism.

The body has an electrical system and we need data about that too. While electricity zips along wires at around 1 millimetre per second (though the associated electromagnetic wave travels at around the speed of light, 300,000 kilometres per second), signals move around our bodies at 0.08 km per second, or about 290 kph. The electricity in our bodies is carried by larger, more complex ions rather than the nimble electrons—charged subatomic particles —that power our homes.

Studies of how impulses ripple down nerves date back to the development of the "voltage clamp" method in the 1930s and 1940s by the biophysicist Kenneth Cole (1900–1984) in the US, along with Alan Hodgkin (1914–1998) and Andrew Huxley (1917–2012) in Britain, who found a way to make measurements by threading electrodes down the giant axon—nerve cell—of a squid.

More insights into the body's "wiring" arose from a technique that allows the registration of minuscule electrical currents of around a picoampere—a millionth of a millionth of an ampere—that pass through a single ion channel, a single molecule or complexes of molecules that allow ions to thread through the membrane of a cell. In 1976, the German cell physiologists Erwin Neher and Bert Sakmann reported how to do this using a tiny yet simple device called a patch clamp electrode.

They used the tip of an extremely fine glass pipette to touch a tiny area, or patch, of the cell's outer membrane that, with luck, contained a single ion channel. Applying a small amount of suction through the pipette clamped a tight seal such that ions could only flow from the channel into the pipette. Using a sensitive electrode, they could record tiny changes in current as ions flowed through the clamped channel. For this remarkable insight, Neher and Sakmann would share a Nobel Prize in 1991.

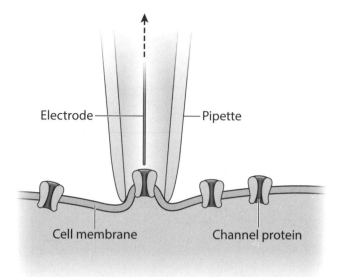

Electrode—————Pipette

Cell membrane　　Channel protein

FIGURE 6. The patch clamp method. Adapted from "A Breakthrough Method That Became Vital to Neuroscience" by Alexander D. Reyes (*Nature* 2019).

But the data that have commanded the most attention in recent years revolve around reading the human genetic code. For that we can thank Briton Fred Sanger (1918–2013), one of the greatest innovators in molecular biology: "Of the three main activities involved in scientific research, thinking, talking, and doing, I much prefer the last and am probably best at it."[11] He was indeed. After becoming the first to reveal the structure of a protein, which happened to be insulin, Sanger would develop DNA-sequencing methods in the mid-1970s, for which he became a Nobel laureate for the second time.

Since Sanger's pioneering work, the cost of sequencing a human genome, the genetic code in a person's DNA, has dropped precipitously—from billions of dollars to hundreds. One reason is the rise of "next-generation" sequencing, an advance that has been likened to the leap from the Wright brothers' aeroplane to a modern Boeing aircraft.

Cambridge University chemists Shankar Balasubramanian and David Klenerman began to develop their next-generation method in 1997, where sample DNA is fragmented into pieces that are

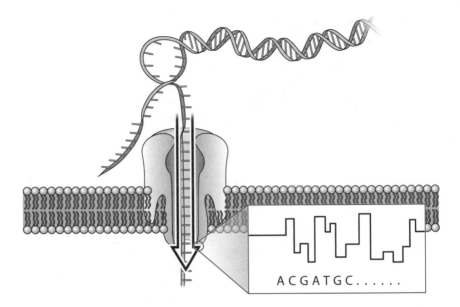

FIGURE 7. How DNA passing through a nanopore channel generates a signal. (Designed by Yoritaka Harazono. TOGO TV. CC BY 4.0)

immobilised on the surface of a chip and locally amplified. Each fragment is then decoded, down to each "letter" of genetic code (a nucleotide—more later), using fluorescently coloured letters added by an enzyme. By detecting the colour-coded letters incorporated at each position on the chip—and repeating this cycle hundreds of times—it is possible to read the sequence of each DNA fragment.[12]

Another advance in next-generation sequencing dates back to the 1970s, when Steve Hladky and Denis Haydon at Cambridge recorded the current flow through a single ion channel in an artificial membrane. Because DNA is a charged molecule, it can also be drawn through this open channel and, as it passes through, causes current fluctuations that correspond to the genetic sequence. The subsequent development of commercial "nanopore sequencing" by Oxford Nanopore Technologies can be traced to research in the 1980s, notably by its founder Hagan Bayley, when the movement of DNA through pore proteins was first observed.[13]

Using this kind of sequencing technology, it is now possible to read DNA stretches that are considerably longer than those managed by earlier sequencers so that, in 2021, an international team of 30 institutions—the Telomere-to-Telomere (T2T) Consortium—published the first "long read" genome.[14] This was significant since the historic draft human DNA sequences published in June 2000 missed as much as 15% of the genome because earlier sequencing technologies read the code of millions of fragments of DNA in parallel, each relatively small, consisting of up to 300 letters of code. As a result, they could not deal with repetitive stretches of DNA code that lurk in the genome, notably the centromeres, the pinched parts of chromosomes that play a key role in cell division. The 2021 end-to-end sequence, based on long reads of between 10,000 and 100,000 letters, revealed 115 new genes that code for proteins, and probably contains many regions that play a role in gene regulation and other functions.

The Code of Life

When details of the entire human genetic code began to swim into view two decades ago, the question of what data it takes to define a human being appeared to have a beguilingly simple answer. Details of the rich complexity of the body seemed to reside in DNA, biology's best-known carrier of information. At the start of this century, the public were led to believe that understanding the code would herald the era of personalised medicine.

Biologists knew that the human genome was a hugely important resource, and of that there can be no doubt. A copy resides in each one of your one hundred trillion cells (with the exception of red blood cells—they destroy their DNA to carry as much oxygen as possible, while staying small enough to squeeze through capillaries). Like volumes in a "Library of You," your DNA is bundled up in packages known as chromosomes. Typically, there are 46 chromosomes in human cells. If you take the largest, chromosome 2, the DNA that it contains would measure more than 8 centimeters when unfurled.[15]

Use X-rays to study the twisted, helical DNA in each of these bundles and you can understand how it carries data. Within DNA's double helix is a ladder of coded information, where each "rung" consists of two chemical units, called nucleotide bases. These units come in four types: adenine, A, thymine, T, guanine, G, and cytosine, C. Because of their shape and chemical properties, the bases always pair up within a rung in the same way: C only pairs with G and A only pairs with T. There are six billion of these letters in our 46 chromosomes.

That is why the double helix also holds the secret of how cells are able to pass on their instructions after they divide: if you peel apart the two strands of the double helix, the rungs of the ladder separate into complementary bases. Each resulting strand can act as a template to copy its original partner strand and thus preserves the information on how to create the proteins that build and run the body (with the help of lots of cellular error correction machinery).

The order in which bases appear spells out the code of life in a similar way to the letters in this sentence, only the messages they carry spell out the instructions to make a protein, one of the building blocks of cells, via the intervention of a related genetic molecule, called RNA. The information in genes is written in a three-letter code, with a triplet of DNA letters—a codon—being responsible for a particular amino acid that, when linked to a string of other amino acids, folds into a protein, one of the units that build and operate your cells.

Even though there are only 20 different amino acids, the cells in your body use a vast array of combinations to build you, from proteins as different as haemoglobin, the red pigment that transports oxygen in your blood; insulin, the signalling molecule that starred in Sanger's Nobel citation; or the enzyme ATP synthase, an energy-converting molecular machine some 200,000 times smaller than a pinhead, which spins 60 times per second to churn out the energy currency of our bodies, a molecule called ATP.

In all, as stated before, there are some 37.2 trillion cells in the body, and though they (with a couple of exceptions, such as red blood cells for the reasons mentioned earlier) contain all the DNA information of an individual, each kind of cell in the adult body only depends on using a particular subset of genes in your genome. In this way, cells can specialise to be one type, from nerve and muscle cells to those that populate organs such as the brain and heart. No wonder so many think that the human genome possesses all the answers when it comes to questions about human biology.

Beyond DNA

The first person to know his own genetic data—and its limitations—
was the genomics pioneer and entrepreneur Craig Venter,[16] who led
the private effort to produce the first draft sequence of the human
genome in 2000. On September 4, 2007, a team led by Sam Levy at
the J. Craig Venter Institute in Rockville, Maryland, finished reading Venter's own genetic code, marking the publication of the first
complete (all six-billion-letter) genome of an individual human.[17]

Roger edited Venter's autobiography, *A Life Decoded*,[18] and can
remember how even Venter was surprised at how little his genome
was able to reveal back then. At that time, nobody knew how to read
genomes with any real acuity; this is one reason why, in a later venture called Human Longevity Inc. (HLI), Venter—like Leroy Hood
and others—not only harvested genomic information but linked it to
phenotypes: patient anatomy, physiology and behaviour, from online
cognitive tests to echocardiograms and gait analysis.[19]

Screening by HLI identified a broad set of complementary age-
related chronic disease risks associated with premature mortality
and strengthened the interpretation of whole genome analysis.[20]
"We probably save at least one person's life every day at the clinic
by discovering a major tumour they did not know they had," Venter
told us, providing some compelling yet anecdotal examples. In his
case, having been declared cancer-free by conventional methods,
HLI screening revealed prostate cancer that was beginning to spread.
Nobelist Ham Smith, Venter's long-term collaborator, was found to
have a significant lung tumour. These timely diagnoses came not
from genetics but by using a powerful 3 Tesla MRI scanner (around
60,000 times stronger than the Earth's magnetic field) with advanced
imaging analysis. Venter told us that, in this scanner, "tumours light
up like lightbulbs."*

The phenotype, that is, observable traits and characteristics of the
body, from eye colour to cancer is a long way from the genotype, or
genetic recipe, of the body. To think that the one kind of data found
in a genome can reveal the details of a person is like trying to deduce

---

* Craig Venter, interview with Peter Coveney and Roger Highfield, December 29, 2021.

the look, taste and mouthfeel of a cake from the recipe. Some things—fruit and currants—are obvious, but many details are far less so.

Though sequencing of the human genome marked the end of almost a century of effort in hunting for protein-coding genes, it underlined how little we understood about the nonprotein coding regulatory elements that bring the genome to life.[21] Of the three billion letters of DNA in the human genome, only around 2% code for the proteins that build and maintain our bodies. Although the last two decades have been a golden age of gene discovery, around 20% of human genes with a vital function remain shrouded in mystery, according to a study of 'unknomics' carried out by Sean Munro of the Laboratory of Molecular Biology in Cambridge and Matthew Freeman of the Dunn School of Pathology, University of Oxford.[22] There is plenty more work left to do beyond understanding our genes. Once upon a time, introns were among the huge tracts of human DNA—around 98%—that were written off as meaningless junk. We now know these noncoding regions of the genome harbour important regulatory elements that determine how gene expression is controlled, but our understanding remains limited.[23]

There has also been a huge surge in interest in the way that genes are used in the body, in a field called epigenetics. Specialisation into different cell types is down to patterns of gene expression, rather than changes in DNA itself. The nurturing environment begins at the chromosome, the bundles of DNA in our cells. Chromosomes are precisely organised, as are the proteins that interact with them, and this organisation seems to be important for the way genes are used.[24] Patterns of gene use can be based on chemical modifications of DNA—for example, decorate a gene with chemical moieties called methyl groups and you turn that gene off—and of histones, tiny proteins that attach to DNA, like beads on a necklace, and play a role in packaging DNA and in regulating the way genes are activated. As a result, the link between genotype and phenotype is not straightforward.

From DNA to Protein

DNA narratives can be convoluted. Traditional genetics research sought a DNA variant linked with a disease. Sometimes this tells a simple story. Examples include the mutations in the gene responsible for a blood clotting factor that cause the inherited bleeding condition haemophilia, sometimes called the royal disease because it figured prominently in European royal families.

Often, however, these correlations tell a complicated story. When it comes to common brain disorders, such as schizophrenia and Alzheimer's disease, a huge amount of information has come from genome-wide association studies (GWAS), where researchers seek to compare genetic sequences of thousands of individuals with a particular trait. The good news is that hundreds of genomic regions can be associated with a person's risk of developing a brain disorder. But sometimes this is a triumph of data over understanding. Explaining *why* a spectrum of genetic variants impacts health remains a challenge.[25] Even if we do link variants with disease, studies of diabetes, for example, have shown that genetics only account for about 10% of the variance seen in the disease, the rest depending on lifestyle and nutrition.[26]

Mapping from data about the genotype to the phenotype is complicated by the fact that our genes are hugely outnumbered by proteins. Genes can be shuffled and used in clever ways. The ability of each gene to code for many proteins is due to a process known as alternative splicing, where bits of code, called introns, are first spliced out and remaining pieces of the gene, known as exons, may or may not be included when creating the protein. In theory, as many as 100 proteins can be produced from a single gene.[27]

Even for coding DNA regions, there is not a straightforward correspondence between the linear DNA code and the 3D shape of proteins in the body, which is crucial for how they work—for instance, to speed a cellular chemical reaction. For a protein with just 100 amino acids, the number of alternate structures that it can adopt in the watery interior of a cell ranges somewhere between 2 to the power of 100 and 10 to the power of 100 possible conformations (shapes). To explore each one would take an eternity, but this one-dimensional code becomes the correct three-dimensional shape—which is crucial to the way it works—with various kinds of help.

One aid to adopting the right shape comes from the incessant molecular movement in cells caused by heat energy. Many of the vital components of the living cell are small enough to be buffeted by the constant pounding of the sea of molecules that surround them

(Brownian motion), which can help a protein jiggle into the most stable shape even when there are millions to trillions of possible configurations.[28] In addition, there are many mechanisms to fine-tune how the body interprets genetic data and turns it into proteins.

Some of the ways that proteins are helped to fold up in our cells can be found in a remarkable molecular machine called the ribosome, consisting of around half a million atoms and measuring around one millionth of an inch across. This machine lies at the crux of two eras of life on Earth, one familiar and one shrouded in mystery: the former consists of today's DNA-based living things and the latter reflects the very first things that multiplied, which are widely thought to have relied on RNA, a delicate yet flexible kind of genetic material that not only stores information but, unlike DNA, can also catalyse chemical reactions. Indeed, the ribosome is a ribozyme, an enzyme made up of RNA folded into an elaborate structure.

Peer deep inside the ribosome, as structural biologists have done, and you can see an ancient core that has been turning instructions into the proteins to build living things for the best part of four billion years. There a rind of proteins has evolved around the central RNA machinery to hone its performance, and this differs depending on the creature—our ribosomes are almost twice as large as those of the bugs that infect us, for example.

The ribosome needs various ingredients to work: first, a messenger RNA molecule, which carries the instructions for making a protein from DNA. To turn this messenger RNA code into protein, the ribosome harnesses a second type of RNA—transfer RNA—which carries the building blocks of proteins, called amino acids.

We now know the atomic details of how the ribosome turns DNA data into flesh and bone, thanks to Nobel Prize–winning X-ray studies by Venki Ramakrishnan in the UK, Ada Yonath in Israel and Thomas Steitz in the US. They revealed that the ribosome is composed of three different RNA molecules and more than 50 different proteins, arranged in two components (called 60S/40S in our cells, and 50S/30S in bacteria)—one is the "brain" that reads genetic code and the other the larger "heart" that makes protein (figure 8). They drift apart and together, as molecular links are forged and broken, to create the proteins that build and run the body.

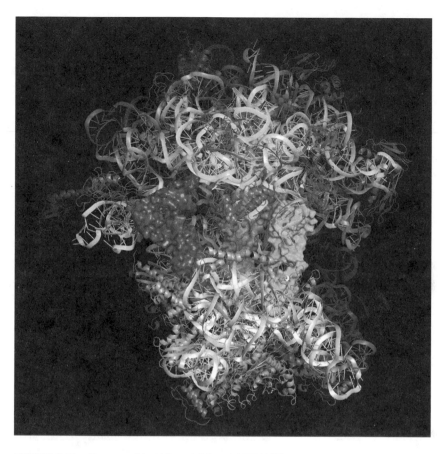

FIGURE 8. The ribosome. (Venki Ramakrishnan, MRC LMB)

These kinds of detailed studies show that the translation of genetic data into protein is not straightforward. Known as gene expression, this process begins with the information contained in the genes (DNA) being turned into a messenger RNA (which also consists of four "letters"—chemicals, called bases—based on the instructions in the DNA, though the base uracil, U, takes the place of thymine, T). Like a tape reader, the ribosome reads the messenger RNA three letters at a time, and matches it to three letters on a transfer RNA, lining up amino acids that they carry in the correct order, then stitches them together. Overall, the ribosome ratchets RNAs laden with amino acids through its core at a rate of 15 per second to string them together into proteins.[29]

Critically, however, the environment once again plays a key role in the resulting configuration adopted by a protein. Gunnar von Heijne at Stockholm University told us how the ribosome helps to nudge proteins into shape. This process, known as cotranslational folding, begins in the exit tunnel of this great molecular machine, which contains nooks and crannies.[30] Pulls and tweaks are provided by docking sites in the exit tunnel that stick to certain parts of the nascent proteins, depending on their sequence of amino acids.[31] There are also "chaperone" proteins that help wrestle many proteins into shape. A handful of different chaperones lurk at the mouth of the exit tunnel, others outside the ribosome. Various quality control mechanisms reject and degrade misfolded proteins before they reach the buzzing chemical chaos of the cell.

Overall, a bewilderingly complex series of steps takes us from a one-dimensional code to a three-dimensional protein building block of the body. Even so, during his acceptance speech for the 1972 Nobel Prize in Chemistry, Christian Anfinsen made a bold claim when he postulated that a protein's amino acid sequence should fully determine its structure in solution. This suggestion became a grand challenge of biology: Can we figure out a protein's three-dimensional structure from its one-dimensional amino acid sequence?

The scale of this challenge had been outlined by Cyrus Levinthal, the American molecular biologist who pioneered the computer graphical display of protein structures. He noted that it would take longer than the age of the known universe to work out the likely configurations of a typical protein by brute force calculation, that is, by testing each and every possible shape it could adopt. Levinthal estimated there were 10 to the power of 300 possible conformations of a typical protein, and it is almost paradoxical that a cell can "work out" the functional version much more quickly than a computer. In Chapter Four, however, we describe how AI has now caught up with cellular reality. When it comes to Virtual You, this feat could offer many insights—for instance, when proteins adopt the wrong shape to cause disease—and help conduct virtual drug trials to find possible treatments.

Colossal Data

We are in the era of what some call "big data." However, when it comes to medicine and biology, big data are actually tiny relative to the complexity of a cell, tissue or organ. Moreover, all these data constantly change in living things. Data scientists like to talk about "the three Vs": volume—amount of data; variety—complexity of data and the sources that it is gathered from; and velocity—rate of data and information flow. Today, the velocity, variety and volume of data all seem overwhelming, though we are a long way short of capturing the full complexity of the human body.

Others talk about the veracity of data, which is also changing. Most traditional data are structured data, that is, neat enough to be directly pasted into spreadsheets and structured databases. Think of a traditional laboratory notebook with one quantity, say, the colour of reactants in a test tube, shown in one column and in another the pH, or the number of offspring from a rabbit population, plotted against time.

When it comes to building Virtual You, there are all sorts of structured data sets that are routinely gathered about patients in the form of the results of temperature, blood and urine tests. There are increasingly extensive genetic data too, as the cost of sequencing our DNA has plummeted. At the molecular level, there is geometric information on the shape of sites on proteins, where other molecules interact, along with scalar and vector fields, used by mathematicians and scientists to show quantities that change—for instance, the concentration of a biochemical across a cell. You can also summon information about proteins, thanks to proteomics, and about metabolism too, using metabolomics. A few Christmases ago, Roger selflessly worked with a team at Imperial College London to study the metabolic impact of hangovers.[32] The suffering seems to be linked to an unusual profile of sugar-like molecules and alcohols, polyols, which play a role in dehydration.[33]

But we have now also entered the era of "unstructured data" as the internet digitizes anything and everything mediated by a microchip, from tweets and text messages to social media posts and YouTube uploads. Around the world data are pouring out of sensors, mobile phones, apps and indeed just about anything you can think of as chips become embedded in humdrum domestic items, from printers to fridges, to form the so-called internet of things.

Increasingly, smartphones and other wireless devices will be used to gather patient data—for instance, on movement, activity patterns, blood pressure, heart rate, calls placed and received, keyboard use, and natural language processing—for "digital phenotyping," which can in turn be used for things like monitoring bipolar disorder, detecting problem drinking or recognising distress.[34] While the diagnosis of mental health was once purely subjective, it can today be made objective with the help of a wealth of personal data about activity and speech patterns, tone and breathing—from laughs to sighs.

FIGURE 9. A detailed high-resolution anatomical model created from magnetic resonance image data. (Credit: IT'IS Foundation)

In the long run, according to our UCL colleague Andrea Townsend-Nicholson, precise and structured data from the medical community will be calibrated in a way to compare different people, while we will use unstructured data and "lifelogging" from smartphones and other devices to fine-tune the look and behaviour of Virtual You.

Data Integrity

When digital twins become established, each one will mark a symbiotic relationship between a person and their digital double, nourishing each other with data and insights. Practical issues will arise, many of which are familiar. Some scientists are unreliable custodians of these data, even though they increasingly worship them. The research that appears in journals is limited to headline conclusions or summaries of the key findings or dominated by studies that have produced encouraging results, with the rest quietly ignored and left unpublished. The raw data—including the "negative data" from unsuccessful experiments—is often omitted, and is lost to the scientific community, and to future researchers too. Fortunately, this is now starting to change.

Some fear that data will also be lost to future generations because of the use of ephemeral recording media, soon-to-be-obsolete storage devices, and software developed by companies whose business models depend on planned obsolescence and compulsory upgrades. There are many solutions under consideration, from public archives to harnessing a storage medium that has been around for billions of years: a single gram of DNA is capable of storing 215 petabytes (215 million gigabytes) so a container about the weight and size of a couple of pickup trucks could, in principle, store every bit of datum ever recorded by humanity.[35]

We also need consistency, so that we can be sure that different scientists and engineers in different labs will be able to measure the same things, use the same jargon and report their findings in a way that can be interpreted by all, ensuring that science remains reproducible. Ask for another researcher's materials, and it is not unusual to endure a long delay, be ignored or even denied. It can be that there is just not the time, the money, the opportunity or the inclination (for example, if software formats have to be modified). Sometimes too, ego, competitiveness and emotion can get in the way—yes, even in scientific disciplines supposedly grounded in objectivity and reason. As the old joke regarding data mining goes—"the data are mine, mine, mine."[36]

There are tensions here that need to be understood. There is always a trade-off between encouraging widespread use of data and methods on the one hand and commercialisation on the other, not least the use of patents to protect ideas and make revenue. Moreover, researchers are reluctant to share their groundbreaking data if they risk not being the first to publish the critical results and insights that these data contain. Without getting the credit for making sense of the data that they have laboriously harvested, they could lose money and recognition, stymie their career, even do themselves out of a Nobel Prize.

We need openness, so that these data are freely shared in repositories and databanks, whether state run or commercial.[37] Fortunately, there are many notable examples of data being shared for public use, and public good, such as the US NCI Cancer Research Data Commons, the UK Biobank, and the GISAID Initiative to share data from all influenza viruses and the coronavirus causing COVID-19. The good news is that many now adopt what are called the FAIR principles—data should be findable, accessible, interoperable and reproducible.[38]

There are many thorny issues surrounding data. How can people access their own medical data? Do they even own it themselves, or is it the property of the hospitals in which it resides? Do they want access to it, and to manage it? Will health services be willing to hand it out? What about combining it with other people's data for the purpose of clinical and other trials? What about cybersecurity? We need high-level principles, along with a stewardship body, to shape all forms of data governance and ensure trustworthiness and trust in the management and use of our data.[39]

From Data to Wisdom

The first step towards Virtual You depends on data, but they are not enough on their own to understand the link between genetic makeup, environment and phenotype of the human body. We need to understand the networks that link all these data before we can begin to close the loop to turn data into predictive, quantitative biology. Above all else, we need theory.

Theories have explanatory power. They make sense of experimental observations in terms of a deeper understanding of the way the world works. They represent nature's workings in an economical manner, without voluminous repositories of raw and uninterpreted data, and help us uncover principles along with laws that explain how and why things are the way they are. They reveal emergent properties too.

Such laws and theories have a mathematical form; for this is the only way in which we can make logically correct analyses of scientific data that are themselves valid. The idea that mathematics can capture something of the way the world works, even create a Virtual You, dates back to antiquity. Pythagoras (570–495 BCE) declared that "all is number." Aristotle (384–322 BCE) described in his *Metaphysics* how the Pythagoreans were so besotted with mathematics that they believed that "the principles of mathematics were the principles of all things."

The better our understanding of a scientific field, and the more powerful our theories, the less we need to rely on amassing hoards of data as diligently as a philatelist would build a stamp collection. Our theories, and the models we build based upon them, are a compressed and flexible way to represent our understanding of nature. The ultimate test of our theories is not how well they agree with experiments after they have been performed, but how well they can predict the result of an observation before it is made.

This is the scientific method at its best, and some of our greatest theories compellingly illustrate how to turn data into insights: among these we can mention the discovery of gravitational waves, one hundred years after Albert Einstein (1879–1955) predicted them—ripples in space-time—which are generated by violent cosmic events, such as when two black holes collide. Another example is the 2012 discovery of the Higgs boson, some half a century after the theorists Peter Higgs, Robert Brout and François Englert proposed that this fundamental particle was associated with the Higgs field, a field that gives mass to other fundamental particles, such as electrons.

Biology needs big ideas too. As the Nobelist Paul Nurse has put it, "More theory is needed. My exemplars for this include the evolutionary biologists Bill Hamilton and John Maynard Smith, and the geneticists Barbara McClintock and Francis Crick. Their papers are permeated with richly informed biological intuition, which makes them a delight to read. This sort of thinking will accelerate a shift from description to knowledge."[40]

Theory makes sense of data, along with how they are interpreted. It does not bode well for the fidelity of Virtual You if we break the laws of thermodynamics, or violate the conservation of mass, momentum and energy. Above all else, theory helps identify essential data—those data that can be used to predict how the body will behave when fed into a computer model. This enables us to focus on emergent properties, so that we avoid the curse of Borges's cartographers' guild. Once we have gathered enough data about the body, the second step towards Virtual You is to convert this raw information into mathematical understanding. We need to create a mathematical model of the human body.

2

Beyond Bacon's Ants, Spiders and Bees

Those who have handled sciences have been either men of experiment or men of dogmas. The men of experiment are like the ant, they only collect and use; the reasoners resemble spiders, who make cobwebs out of their own substance. But the bee takes a middle course: it gathers its material from the flowers of the garden and of the field, but transforms and digests it by a power of its own.

—Francis Bacon, *Novum Organum, sive Indicia Vera de Interpretatione Naturae* (1620)[1]

Curiosity is perhaps the most fundamental of all human traits and, of all the topics that pique our interest, few surpass speculating about what is going to happen next, particularly when it comes to our own destiny. The ultimate aim of Virtual You is to enable doctors to gaze into the future of a patient, not a typical patient or an average patient but a particular patient, with their individual baggage of inheritance, upbringing and environment.

Our remarkable ability to anticipate what is about to happen next is the second ingredient of Virtual You and dates back at least to around the eighth century BCE, when the Babylonians turned systematic observations of the night sky into predictions of the positions of the Sun, the Moon and the known planets.[2] The Scientific Revolution that followed the Renaissance turned that powerful and perspicacious blend of observations with understanding into the scientific method, arguably the greatest achievement of our species. By the twentieth century, we had the basic ingredients to create Virtual You once scientists had worked out how to mechanise much of mathematics with computers, the subject of the next chapter.

When it comes to showing how theory and experiment work together, the philosopher and statesman Francis Bacon (1561–1626) was among the first to articulate the scientific method. We became fascinated with Bacon's prescient writings a few years ago when—with our American colleague Ed Dougherty—we explored the relationship between theory, experiment, data and artificial intelligence in a paper for the *Philosophical Transactions of the Royal Society*,[3] the world's first and longest-running scientific journal, itself born of the Scientific Revolution, in 1665.

Bacon couched the efforts of his peers to understand the world, whether the workings of an organ or movements of the heavens, in terms of the spiders and the ants, where the former use reason and the latter rely on experiment. Importantly, Bacon realised that a blend of both skills—as performed by the bees—was necessary for what today we call science: like these busy little insects, worker scientists harvest data from experiments so that their minds can transform them into nourishing information. In his evocative prose, one can see an outline of how science still works: reason guides experiment, which paves the way for new reasoning and then new experiments. With his metaphor of the bee, Bacon had rejected the approaches of the ant, which depended on radical empiricism (data without reason), and of the spider, which relied on rationalism (reason without data), in his quest for knowledge.

There are endless examples of the buzz created by Bacon's bees. Without the painstaking astronomical observations of the Danish nobleman Tycho Brahe (1546–1601), the German astronomer and mathematician Johannes Kepler (1571–1630) would not have determined that the planets move in elliptical orbits. The most famous prediction of the theory of general relativity—the bending of starlight by a massive body—was confirmed by measurements made by Arthur Eddington (1882–1944) during the 1919 solar eclipse, making a superstar of the author of that theory, Albert Einstein (1879–1955).[4] In May 1952, while a graduate student under the supervision of Rosalind Franklin (1920–1958) at King's College London, Ray Gosling (1926–2015) took Photo 51, an X-ray diffraction image of DNA. Without this striking X-shaped pattern of scattered X-rays, Watson and Crick would not have deduced the double helix structure of DNA, a magic key for unlocking genetics. However, we should emphasise, as historians of science always do, that the reality of scientific advancement is always more complex than heroic narratives suggest.[5]

Scientific Reason

When modern scientists, notably physicists, talk about the approach of Bacon's spiders—"reason"—they mean more than expressing our understanding in a rational, logical way, or deductive and inductive approaches to reasoning, the former working from the more general to the more specific and the latter conversely. By reasoning, scientists mean spinning a web of understanding in the form of supposedly cast-iron mathematical logic, based on equations.

The second step towards Virtual You is to use one or more equations to capture the details of how reality works, whether the electrical activity of a nerve cell, the force generated by a muscle, or the rhythms of a diseased heart. What is behind these mathematical formulae? We do not know and it does not matter. All that really counts is that these mathematical expressions capture a facet of reality in the sense that they describe objective observations.

With regard to the tug of gravity that means our blood pressure is higher in our feet compared with our heads, for example, something certainly exists to cause a blood pressure gradient, but its nature and substance are beyond our ken. The fundamental principle of *Hypotheses non fingo*—"I frame no hypotheses"—is summarised in one of the most famous books of all science, Sir Isaac Newton's *Philosophiæ Naturalis Principia Mathematica* (1687).[6]

This powerful idea lives on. Roger once asked Stephen Hawking (1942–2018) about the nature of imaginary time. Hawking replied: "All we can ask is whether the predictions of the model are confirmed by observation. Models of quantum theory use imaginary numbers, and imaginary time in a fundamental way. These models are confirmed by many observations. So imaginary time is as real as anything else in physics. I just find it difficult to imagine."[7]

From Calculus to Computers

The reasoning power of mathematics has evolved down the millennia, but its ability to describe the world accelerated a few decades after Bacon died with the invention of calculus, a tool to study change, not least that within the human body. It was developed independently in England by Isaac Newton (1643–1727), who featured it in his *Principia*, and in Germany by Gottfried Wilhelm Leibniz (1646–1716). These huge figures became embroiled in a *Prioritätsstreit*, a "priority dispute," even though elements of calculus had been glimpsed long before in ancient Greece, Egypt, India and the Middle East.

Calculus comes in two parts: differential calculus, which focuses on rates of change that can be seen in the slope of a curve, such as miles per hour; and integral calculus, where the accumulation of quantities, such as the overall number of miles travelled, is calculated by working out the areas under or between curves (figure 10). The resulting branches—differential and integral—are united in calculus, which tackles both by breaking them up into very small (infinitesimal) pieces. While differential calculus divides things up into tiny pieces and tells us how they change from one moment to the next, integral calculus joins (integrates) tiny pieces to reveal the overall result of a series of changes. The former produces derivatives, which measure instantaneous change at a single moment in time, and the latter produces integrals, which piece together an infinite number of tiny pieces that add up to a whole, such as an infinite stream of moments. In this way, calculus converts our ineffable world of flux into precise mathematical expressions.

Calculus does tend to instil a degree of fear in many people, but as the physics professor Silvanus Thompson (1851–1916) wrote more than a century ago in *Calculus Made Easy*, "The fools who write the textbooks of advanced mathematics—and they are mostly clever fools—seldom take the trouble to show you how easy the easy calculations are."[8] He concluded that they offer the keys to "an enchanted land."

When it comes to taking the second step towards Virtual You, for example, calculus can unlock the cause of change in the human body, from heartbeats[9] to fluctuating metabolites. The science biographer and populariser Graham Farmelo likened the mathematical short-

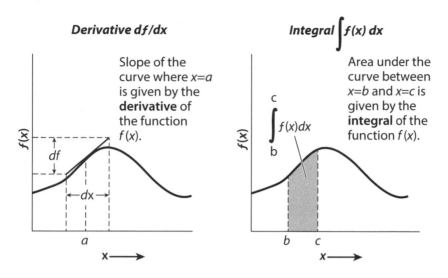

FIGURE 10. Image showing calculus, derivatives and integrals.

hand of differential equations to poetry itself: "If 'poetry is language in orbit,' as the Nobel laureate Seamus Heaney later observed, then differential equations are mathematical language in orbit."

Different classes of these equations demand different methods and varying degrees of exertion to solve them. In general, they can be divided into linear and nonlinear differential equations. The linear variety links quantities in a straightforward way, for instance, as X rises in proportion to Y (to the power of 1). The number of oranges you can buy from a supermarket is in proportion to how much money you spend on them: double the money buys you double the fruit. The nonlinear differential equations describe a more complex relationship, so X rises in proportion to Y to another power, whether 2 or ½ or 100. In a typical (nonlinear) supermarket, the more oranges you buy, the bigger the discount. Nonlinearity can also lead to unexpected effects: with enough oranges you can open a marmalade factory.

At its very simplest, linear equations are simple to solve (you only need a pen and paper) while the outcomes of nonlinear ones are complicated. The latter mean much more than being curvy: they have solutions that are not smooth, can show discontinuities, and settle on multiple stable states instead of only one, so that more than one answer is possible. These hard-to-solve nonlinear equations—

which must be tackled with clever mathematics and computers—also best capture the mathematics of twists, undulations, arcs and bends. They are tailor-made for the human body.

However, our brains crave straightforward narratives. We all want a simpler life. Research in cognitive psychology suggests that the human mind does not sufficiently respect nonlinear relationships.[10] Nor is the pervasiveness of nonlinear phenomena reflected in research for depressingly practical reasons: linear mathematics is much easier to solve. One can even see this in the word *nonlinear* itself because the exception really is the rule. The Polish-American scientist Stan Ulam, whom we encounter in the next chapter, once joked that it is a bit like calling zoology the study of "non-elephant animals."[11]

The ubiquity of nonlinearity was revealed in 1889 by the French polymath Henri Poincaré, who was famously dubbed *un monstre de mathématiques* by his teacher. Poincaré found that as soon as one tries to analyse the motion of as few as three bodies, such as the Sun, the Moon and the Earth, an exact solution cannot be found by "pen and paper" analytical methods (it is called an "intrinsically nonintegrable, nonlinear system").

Poincaré had discovered what we today call deterministic chaos, where equations produce predictable results—they are deterministic— but are exquisitely sensitive to the numbers fed into them. Without infinite precision, there is a limit to how far the future evolution of a chaotic system can be predicted. This limit, known as the Lyapunov time, represents a deep crack in the crystal ball of scientific theory. Its existence is one of the principal reasons why scientific predictions are at best probabilistic.

These nonlinear equations can be further divided into the sub-class of ordinary differential equations, ODEs, which describe continuous changes in a single variable, typically one dimension in space or in time. They are used to chart the motions of the heavens, the rates of chemical reactions in the body and the rise of infections. The ODEs are a subset of a second, bigger, group known as the partial differential equations, PDEs, which deal with how things change as a function of more than one variable, such as time *and* space, the latter being capable of description in several dimensions.

PDEs abound. They pop up in our theories of the very big and the very small—such as general relativity and quantum mechanics—and capture the way we understand heat, diffusion, sound, electromagnetism, fluid dynamics and elasticity. To describe the motion of fluids, for example, we can use a famous set of partial differential equations devised by the French engineer and physicist Claude-Louis Navier (1785–1836) and the Anglo-Irish physicist and mathematician George Gabriel Stokes (1819–1903). Known as the Navier-Stokes equations, they are used in all sorts of digital twins, for instance, in fluid flow in gas turbines,[12] oil and gas reservoirs,[13] and arteries,[14] yet we are still struggling to unlock their mathematical secrets.

However, the effort to put differential equations to work in the virtual human began with research on the electrical state and electrophysiological properties of cells, which have proved a fascination, if not inspiration, for centuries. When Mary Shelley—née Mary Godwin—wrote *Frankenstein* in the early nineteenth century, she knew about gruesome public demonstrations in which electricity was used to reanimate limbs and bodies that had been recovered from the gallows. Just as the crude application of electrodes to animate the dead had entranced the scientists of her day, centuries later the research community would become just as enthralled by studies of an axon, the threadlike part of a nerve cell that conducts impulses, in this case to enable a squid to move.[15] This marked the first step towards the creation of Virtual You.

Atoms to Impulses

Thanks to the conveniently large nerve fibres from the longfin inshore squid, experiments in the 1950s helped to hone a groundbreaking mathematical theory behind the action potential (nerve impulse), which earned Britons Alan Hodgkin and Andrew Huxley the Nobel Prize in 1963. Their theory represented flux and change—the electric current along the nerve produced by the flow of ions—in the form of a nonlinear differential equation, though Hodgkin would joke that their prize should have gone to the squid.[16]

Working at the University of Cambridge, Hodgkin and Huxley showed how, in response to a stimulus, individual nerve cells propagate action potentials, generated by the rapid surge of ions across a nerve cell. Though fleeting—lasting several thousandths of a second—the action potential triggers ion movements in adjacent membranes, like a Mexican wave. By the standard textbook account,[17] the action potential propagates to the nerve terminal, where it triggers the release of a signalling chemical, or neurotransmitter.

Nerves, like other cells, communicate with their environment by the flow of metal ions through channels in their membranes. We now know that the channels themselves are huge and complex proteins that sit inside the cellular membrane and selectively and rapidly transport ions. Hodgkin and Huxley carefully measured currents and linked them to the movement of sodium and potassium ions when potentials (voltages) were applied across the cell membrane.

Treating the membrane as a kind of electrical capacitor, Hodgkin and Huxley were able to use electrical circuit theory to devise a mathematical model, a set of four ordinary differential equations that are coupled (in other words, they share variables), and describe four variables that control how the state of the cell changes over time. These "state variables" consist of the current through the cell membrane along with the potassium and sodium channel activations, and the sodium channel deactivation, all of which can be studied experimentally. In addition, the model contains a set of parameters that can be chosen to produce a best fit with the way the voltage across the cell membrane varies with time. In essence, this is the way all such models are created in biophysics and biochemistry.

Their model captures in an elegant and empirical way the link between the ion channels and the resulting measurable electrical current flowing across the cell membrane, marking the first mathematical description of what is called the excitability of a cell.[18] In his Nobel lecture, Hodgkin remarked how "the equations that we developed proved surprisingly powerful." Their calculations revealed good correspondence between the predicted values and the spiky shape of an action potential.

Not only had Hodgkin and Huxley created a highly successful mathematical model of a complicated biological process, the model was confirmed by subsequent detailed molecular studies of the various transmembrane protein channels that allow ions to shuttle in and out of cells. There is nothing more satisfying in science than finding out that a theory built on experimental observations and mathematics can pave the way to deeper insights. It is also telling that these pioneers were biophysicists, underlining the value of combining the efforts of different disciplines. Through their work, Hodgkin and Huxley unwittingly began to lay the foundations for Virtual You.

The Limits of Computation

A remarkable testament to the power of mathematics is that it can be used to show the limits of what computer simulations—the subject of the third step and our next chapter—are able to do. This revelation came in the wake of a challenge set in Paris in 1900 by David Hilbert (1862–1943), a professor of mathematics in Göttingen, Germany, a small university town that had fostered prodigious mathematical talents, such as Carl Friedrich Gauss (1777–1855), Bernhard Riemann (1826–1866), Emmy Noether (1882–1935) and, of course, Hilbert himself, who listed 23 problems to inspire his peers.

Hilbert would come to seek a limited set of axioms and rules of reasoning from which he could generate all mathematical truth. Step-by-step procedures for carrying out operations by blind application of specified rules are called algorithms, named after the Latinized versions of the name of Muhammad ibn Mūsā al-Khwārizmī, a ninth-century Persian astronomer and mathematician (and the Latin translation of his most famous book title, *Algoritmi de numero Indorum* (Al-Khwārizmī on the Hindu Art of Reckoning)).

These sorts of stepwise procedures should be able to prove things that are true, and would be "complete," in that no truth lies beyond their scope, said Hilbert. They would also have to be mechanical, in other words, so explicit that human subjectivity would not come into play. In this way, Hilbert's mission formed the basis for computability theory—the study of the power *and limitations* of algorithms.

Inspired by Hilbert, researchers delivered a number of unsettling revelations about the foundations of mathematics. At the start of the

1930s, a 25-year-old Austrian-Czechoslovak-American logician, Kurt Gödel (1906–1978), established that some mathematical statements are undecidable, that is, they can never be proved true or false. In a sense, he established what computers can't do.* He demonstrated the inevitability of finding logical paradoxes akin to the statement "This sentence is false." As the English cosmologist John Barrow (1952–2020) remarked, "If we were to define a religion to be a system of thought which contains unprovable statements, so it contains an element of faith, then Gödel has taught us that not only is mathematics a religion but it is the only religion able to prove itself to be one."[19]

A key aspect of Hilbert's programme was his so-called *Entscheidungsproblem* (decision problem), fully formulated in 1928. Hilbert wanted to discover if there existed a definite method—a mechanical process, or algorithm—which could, in principle, be applied to any assertion, and which was guaranteed to produce a correct decision about whether that assertion was true.

In 1936, the disconcerting answer would come from a 24-year-old English mathematician, Alan Turing (1912–1954). His paper—"On Computable Numbers"—is one of the most famous in the history of computing and set out an abstract machine able to tackle Hilbert's problem, akin to an old-fashioned typewriter, equivalent to any machine devised to compute a particular algorithm.[20] This led to the concept of a universal Turing machine, which can carry out any algorithm, just like a modern computer can carry out any program. In this way, Turing gave mathematical substance to what it means to compute.

Tackling the *Entscheidungsproblem* involves devising a computer program that can examine a second program and decide whether the latter would ever terminate or simply "loop" forever. Turing showed that the general problem is uncomputable, beyond any computer program, no matter how complicated (we discuss why in our 1995 book, *Frontiers of Complexity*). Working independently in Princeton in 1936–37, the American logician Alonzo Church[21] arrived at similar conclusions, as did the Polish American Emil Post, while at the

---

* Roger Penrose, in conversation with David Eisenbud, "An Evening with Sir Roger Penrose," Royal Society, June 8, 2022.

College of the City of New York.[22] All these approaches were shown to be equivalent: mathematics cannot be captured in any finite system of axioms.

When Hilbert died in 1943, his gravestone in Göttingen was inscribed with the quote *Wir müssen Wissen, Wir werden Wissen* (We must know, we will know), the words that Hilbert had delivered at the 1930 annual meeting of the Society of German Scientists and Physicians in response to the Latin maxim *Ignoramus et ignorabimus* (We do not know, we shall not know). Yet the Church-Turing thesis is more in the latter's spirit: there exist aspects of our world that are shrouded from simulation by computers because they are noncomputable, or nonalgorithmic.

Computing Reality

Virtual You depends on computers being able to calculate the processes at work in the body. However, Turing taught us that there are noncomputable processes. We do not know the extent to which nature is noncomputable, but we do know that examples lurk within physics. We described one in *Frontiers of Complexity*, while discussing the work of Marian Pour-El and Ian Richards at the University of Minnesota. They showed that there exist noncomputable solutions to well-known equations of mathematical physics,[23] notably the wave equation, which governs the propagation of light in space and time—that is, electromagnetic radiation—but applies to all manner of waves, from those that crash on a beach to sounds.

To represent a wave in mathematical form requires a partial differential equation that is second order in both space and time, where the second derivative in time, the acceleration of the wave's amplitude, is linked to its curvature, which is the second derivative of the amplitude in space. You can derive this wave equation from the work of James Clerk Maxwell (1831–1879), the hugely influential Scottish mathematical physicist—and inspiration for Einstein—who revolutionised our understanding of electric fields, magnetic fields, and the propagation of light, X-rays and other electromagnetic radiation. What Pour-El and Richards showed was that, for some classes of computable initial conditions,* the solutions of the wave equation at later times are uncomputable.[24]

However, though off limits to a *digital* computer, these solutions could be computable with a simple kind of *analogue* computer, one that relies on continuously variable real-world properties—not the limited strings of 1s and 0s of floating-point arithmetic in digital computers. For example, to add 5 and 9, simple analogue computers could add voltages, pieces of wood or amounts of fluid that correspond to those numbers, and then instantly obtain the answer of 14. When it comes to computing wave equations for light, analogue computing is familiar to everyone who has worn glasses or used a microscope or telescope: it can be done by a lens.

Consider a classical electromagnetic field that obeys Maxwell's equations. For some classes of computable initial conditions, Pour-El and Richards showed that the field can be focused so that its intensity at the point of focus can be measured by a simple analogue device, even though it is not computable by any digital computer. That does not mean to say a digital computer will not provide an answer. It will, but there is no way of assessing how close it is to the right answer without using an analogue device or finding an exact solution mathematically.

There may be noncomputable processes in biology too. Consciousness itself does not compute, according to the British Nobelist and mathematician Roger Penrose: we ourselves are able to infer the truth or falsity of statements that lie beyond the reach of algorithms—our thinking contains nonalgorithmic and noncomputable ingredients.[25] This remarkable insight, published in his 1989 book, *The Emperor's New Mind*, has been quietly ignored in the scramble to develop general artificial intelligence.

A complementary insight into the limits of algorithmic thinking has come from Stuart Kauffman and colleagues of the Institute for Systems Biology in Seattle.[26] Humans can use situational reasoning—they adopt perspectives, choose goals and can cope with information that is irrelevant, incomplete, ambiguous and/or contradictory.

---

* Technically, they are said to be not continuous—that is, smooth or regular—beyond being once differentiable, and belong to a category of "weak solutions."

They are able to see "affordances," that is, the possibilities in their surroundings. People have an implicit understanding of the purpose that a thing can have as part of the way they see or experience it. An engine block might help propel a car, but a person would know that it can also be jury-rigged to be a "bizarre (but effective) paper weight."

The team argues that a key difference between rule-driven, Turing-complete machines and us is that algorithms cannot identify and exploit new affordances. One reason AI is successful is that it is designed for a given task or target. But, as Gödel found with mathematics, the team argues there will always be aspects of human intelligence that cannot be captured by a formal, rule-based model: "Identifying and leveraging affordances goes beyond algorithmic computation." To compete with humans, artificial general intelligence needs more than induction, deduction and abduction and "would have to know what it wants, which presupposes that it must be capable of wanting something in the first place." When it comes to fears that the rise of general AI could threaten our very existence, they reach a comforting conclusion: "No machine will want to replace us, since no machine will want anything, at least not in the current algorithmic frame of defining a machine."

The Problem with Digital Computers

There is another basic issue with digital computers raised by the work of Turing that stems from the numbers they manipulate. What Turing referred to as the "computable numbers" are those that can be handled by a Turing machine. They include the rational numbers, everyday numbers that can be seen in price tags, shoe sizes, recipes and more, such as the integers (whole numbers, including negative numbers) and fractions that can be expressed as ratios of integers.

But there is also the class of irrational numbers, those numbers that cannot be expressed as the ratio of two integers. Examples of irrational numbers include π, the circumference of a circle with a diameter of 1 unit, and e, Euler's number, which is used in logarithms and to work out compound interest and is ubiquitous in science and engineering. Although some irrational numbers—for instance, π or the square root of 2—can be computed to any degree of precision by an algorithm, the overwhelming majority of irrational numbers are not computable.

This is problematic because, in the sea of all possible numbers, the number of computable rational numbers is vanishingly small. Yes, there are infinitely many rational numbers, which are "countable," but we have known since the time of the German mathematician Georg Cantor (1845–1918) that the set of irrational numbers is infinite in a bigger way, being "uncountably infinite."

Cantor's discoveries encountered hostility from his peers, even from Christian theologians who saw them as a challenge to God. In contrast, Cantor (like others before and since) thought he had actually glimpsed the mind of God.[27] Now long accepted, his work indicates that, by manipulating only the computable numbers, digital computers may neglect the richer and deeper possibilities raised by using the infinitely bigger palate of noncomputable numbers. When you take a close look at the numbers that inhabit digital computers, there are further shocks in store.

Flops

The current effort to simulate Virtual You rests on conventional, or classical, computers within which mathematical operations—such as addition or subtraction—are conducted on strings of 1s and 0s of limited length. These are known as floating-point numbers. They are all rational numbers, flexible, hence the "floating point"—and were designed to allow computers to manipulate numbers of wildly differing magnitudes that range from the span of a spiral galaxy over hundreds of thousands of light-years to the height of a person and to the dimensions of cellular ion channels a few nanometres across. While a difference or error in measurement of a few billionths of a metre matters little when it comes to the span of a galaxy, it can make a huge difference when it comes to the molecules of life.

These flexible floating-point numbers have three parts: the digits, known as the significand or mantissa; the exponent, which shows where the decimal point is placed; and the sign bit, which is 0 for a

Single precision

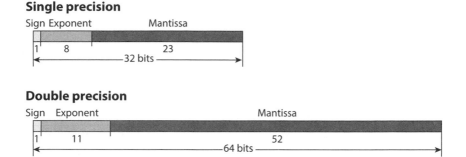

Double precision

FIGURE 11. Single and double precision floating-point numbers.

positive number and 1 for a negative number. In a computer, these are expressed in binary so that the significand consists of 23 bits, the exponent of 8 bits and the sign bit (figure 11). Most modern computers are fuelled by these "single precision IEEE floating-point numbers," which refers to a technical standard established by the Institute of Electrical and Electronics Engineers in the 1950s. That is also why the power of a computer is measured in terms of flops, which refers to floating-point operations per second.

While these floating-point numbers consist of 32 bits in all, and provide 2 to the power of 32 numbers (about four billion in total), there are also double precision numbers that manipulate 64 bits and offer vastly more numbers, at 2 to the power of 64. Even higher levels of precision exist and there are also half precision numbers, with just 16 bits, which have become popular in machine learning to overcome memory limitations and boost speed.

Because computers rely on this decades-old floating-point arithmetic, they are hugely inefficient. John Gustafson of the National University of Singapore/A*STAR likes to jest that IEEE floating-point numbers are "weapons of math's destruction."[28] He points out that if you organised your working day in the way a computer does, which puts a disproportionate amount of effort into moving data around, you would commute for four hours, work for five minutes, then spend another four hours travelling home. As a consequence, 0.2 billionths of a joule of energy are expended in adding 64-bit numbers, while 12 billionths of a joule are expended reading 64 bits from memory.

To overcome this "memory wall," and move only meaningful data, Gustafson has developed the universal number, or unum format, which gives more accurate answers than floating-point arithmetic yet uses fewer bits in many cases, saving memory, bandwidth, energy and power.[29] Gustafson has also proposed a hardware-friendly version of unum, an outgrowth of the floating-point number format, called "posits": "If an algorithm has survived the test of time as stable and "good enough" using floats, then it will run even better with posits."

For the time being, however, computers are stuck with a limited repertoire of rational, floating-point numbers. How does this limit our ability to simulate the world?

Lost in Translation

Before we can simulate the human body in a computer, our model has to be fed relevant data. To represent flows of blood, for example, we have to turn the smooth, continuously changing pressures and flow velocities into the 1s and 0s of floating-point numbers (figure 12).

When it comes to the single precision numbers, for example, all the messiness of reality has to be boiled down to four billion numbers that range from plus to minus "infinity" (not actually infinity, of course, but just "very big"—roughly 2 to the power of −127 or +127). You can capture more of this reality by using arrays of numbers, or by using double precision, which offers 16 quintillion different numbers, but the floating-point numbers are not evenly distributed over this vast range. Far from it.

Single precision numbers are exponentially distributed as decreasing powers of 2. There are as many of these numbers grazing between 0.125 and 0.25, as there are between 0.25 and 0.5, or between 0.5 and 1, and so on. Of the four billion, approximately one billion lurk between 0 and 1, another billion between 0 and −1, while the remaining two billion cover from ±1 to ±"infinity" (figure 13).

The way that this diminished set of numbers can cloud the crystal ball of computer simulation was first laid bare in the early 1960s at the Massachusetts Institute of Technology in Cambridge, Massachusetts, when the meteorologist Edward Lorenz (1917–2008) used

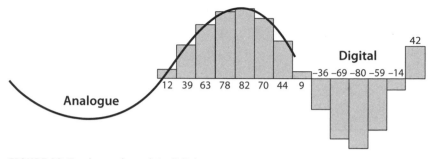

FIGURE 12. Turning analogue into digital.

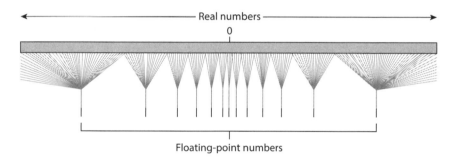

FIGURE 13. Floating-point numbers capture only a tiny fraction of mathematical reality.

a desk-sized LGP-30, or Librascope General Purpose computer, to develop a simple forecasting model.

Lorenz was trying to simulate the Earth's atmosphere with a set of three coupled ordinary differential equations—simplified Navier-Stokes equations—that describe convection, the fluid movement you can see in pans of boiling water or a cup of hot coffee. But when he tried to repeat his initial simulation, he got different results. On closer inspection, Lorenz found that the differences resulted from a tiny rounding error in the numbers fed into his computer.

You might expect that a small error would lead to a small difference. The reality was shocking: wildly diverging results span out of a change in even the last decimal place in his numbers. Yet these apparently random fluctuations arise from deterministic laws, governed by differential equations, which means the outcome is determined entirely by the numbers you feed into them.

His observations about the sensitivity of chaotic systems ultimately led Lorenz to formulate what became known as "the butterfly effect"—reflected in the title of a paper that he presented in 1972: "Predictability: Does the Flap of a Butterfly's Wings in Brazil Set Off a Tornado in Texas?"[30] Ever since, efforts have been under way to investigate, grasp and tame deterministic chaos.

From Attractors to Ensembles

A mathematical object that is as useful as it is beautiful has become an icon of chaos, a means to understand it, even to exploit it too. In his model of convection, Lorenz had identified three key influences—the temperature variation from top to bottom, and from side to side, and the speed of convection. He could plot these properties as he tracked them moment by moment.

Working with Ellen Fetter, Lorenz found the hallmark of chaos. Just as the weather has regularities, manifested in the form of the seasons from summer to winter, yet varies hugely day to day (it certainly does in London), so they found that the point would trace out a butterfly-shaped trajectory overall yet would never follow the same path around the "wings."

Begin with minutely different starting points and you always end up on different tracks as the computer steps forward in time: two points on the attractor that initially lie next to each other end up being exponentially far apart. However, whatever the starting point, you end up trapped within the overall butterfly shape, which is formed of many trajectories.[31]

To understand the full scope of what may happen in a chaotic dynamical system, one must probe the behaviour on this butterfly shape for all—or as many as possible—initial conditions. The resulting set of simulations starting from a constellation of initial conditions is called an ensemble, an approach that aligns closely with the one used by Boltzmann to calculate how temperature and other bulk, or macroscopic properties, emerge from the microscopic—molecular—description of matter; say, the velocities of water molecules, which gave rise to an entire field, known as statistical mechanics.[32]

In this case, the ensemble is generated by simulating the entire system of water molecules—say, a kettle of water—and then repeating this a vast number of times, using different initial conditions. Although you can compute the temperature from the kinetic energy—that linked with movement—of all the different water molecules in each "replica" of a kettle full of water, the thermodynamic temperature will be the ensemble average, namely, the average over each and every replica, an idea introduced by the American J. Willard Gibbs (1838–1903).

Ensembles also featured in some of the first simulations of atomic behaviour used to model nuclear chain reactions, which we return to in the next chapter. Today, they are employed extensively, notably in weather forecasting, where slightly different starting conditions and values of uncertain parameters are fed into weather prediction models to cover the range of possible expected weather events—for instance, where a hurricane might strike. One run of a code that is exquisitely sensitive to its initial conditions will produce different answers from the next. Importantly, however, these ensemble calculations do generate seemingly reliable averages.

The English physicist Tim Palmer, working with James Murphy at the UK's Meteorological Office, produced the world's first probabilistic

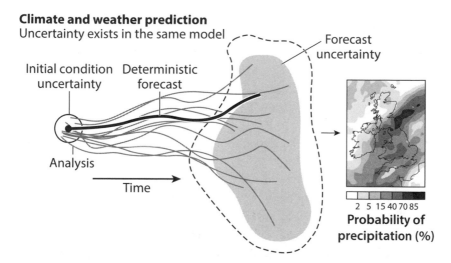

Climate and weather prediction
Uncertainty exists in the same model

Forecast uncertainty

Initial condition uncertainty   Deterministic forecast

Analysis

Time

2 5 15 40 70 85
Probability of precipitation (%)

FIGURE 14. Ensemble forecasting. Adapted from "The Quiet Revolution of Numerical Weather Prediction" by Peter Bauer, Alan Thorpe and Gilbert Brunet (*Nature* 2015).

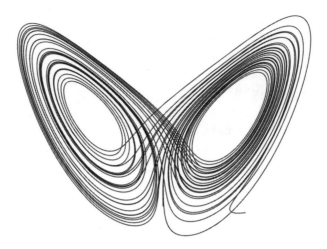

FIGURE 15. The Lorenz attractor. (Shunzhou Wan, Centre for Computational Science, UCL)

weather forecast based on ensembles over 25 years ago.[33] They found that if the forecasts vary hugely across the ensemble, overall predictions have to be hedged in terms of significant uncertainties, whereas if they are very similar, confidence rises.

In this way, chaos can be quantified. By calculating the properties of an ensemble, we move from an attempted but unreliable deterministic description to a reliable but probabilistic one, one couched in terms of averages and standard deviations. However, mathematics also shows us why chaos can never be fully tamed.

Strange Attractors

The butterfly-shaped mathematical picture we described above is a Lorenz attractor, where the term *attractor* means a state to which a dynamical system evolves regardless of its starting state, and where a dynamical system means anything that changes in time, from a swinging pendulum to a human being, or indeed a population growing according to a specified mathematical relationship. When it comes to driven (that is, far from equilibrium) dynamical systems, for example, a ball that rolls to rest on the bottom of a bowl is said to have reached a stable equilibrium, called a fixed point attractor.

Meanwhile, a heart cell's rhythms are captured by what is known as a periodic attractor. Because Lorenz's butterfly wings never settle on either of these kinds of attractors, however, they go by the name of a strange (or chaotic) attractor.

The Belgian mathematical physicist David Ruelle first coined this evocative moniker with the Dutch mathematician Floris Takens.[34] A strange attractor has what is known as fractal geometry, a term famously conceived by the French-American mathematician Benoit Mandelbrot (1924–2010), which has the property of self-similarity. Zoom in on the tracks of the strange attractor, and you will see that they follow similar, but not identical, trajectories and orbits. Zoom in once more, and you are presented with the same picture. And so on to infinity.

Familiar examples of fractal objects are cauliflowers or Romanesco broccoli, where, when you zoom in on a floret, you see a smaller version of the same thing. Fractals lurk within us too, from fluctuations in the rhythm of a healthy heart to the distribution of blood in the circulatory system, the structure of airways in the lungs, and of neurons too.[35]

The work of Lorenz also opened up an alternative way to sum up the chaos of turbulence in the body because a strange attractor captures the chaotic essence of a flowing fluid, albeit requiring infinite dimensions rather than just three. Just as a complex work of music can be broken down into pure notes, so turbulence can be expressed in terms of an infinite number of periodic orbits. Turbulence may seem disordered, but the strange attractor suggests instead that a fluid explores this fractal nest of orbits. Despite being called an attractor, it actually consists of orbits that the fluid never sticks to for long: the orbits are unstable, being at once attracting and repulsive in nature, so that in turbulent flows the fluid moves seemingly randomly from one orbit to another. As a result, these unstable periodic orbits can be thought of as the fundamental units of turbulence, rather like simple vibrations are the fundamental units of any notes played by a musical instrument. Predrag Cvitanović of the Georgia Institute of Technology, Atlanta, calls these orbits the "skeleton of chaos."[36] In the next chapter, we will see how Peter has used this as a skeleton key to unlock turbulence.

Digital Pathology

Another problem still nagged at Peter when it came to using computable numbers for a classic so-called multiscale problem of the kind tackled by Boltzmann, like working out the "macroscopic" properties of a substance such as a gas—pressure, temperature and so on—from its molecular properties. While mulling this over with Shunzhou Wan at University College London, Peter realised computable numbers might lead to problems for exactly the same reason that rounding numbers caused a headache for Lorenz.[37]

Working with the mathematicians Bruce Boghosian and Hongyan Wang at Tufts University, Boston, Peter decided to investigate. They explored a basic example of chaos, called the generalised Bernoulli map, with floating-point numbers. Though simple, the map is mathematically equivalent to many other dynamical systems encountered across physics, chemistry, biology and engineering. Biologists already knew that, despite the simplicity of Bernoulli and other maps, they were beyond normal computer simulation because of the limited precision of computers.[38]

But the full extent of the problem with floating-point numbers had not been explored until Peter and his colleagues fed the map with each and every possible initial condition in the interval between 0 and 1—all one billion single precision floating-point numbers—and compared ensembles of their simulations against the "right" answer (the Bernoulli map is unusually amenable to complete analysis using mathematical tools that rule the real, continuum world, such as integration, differentiation and power series expansions). What they found was surprising.[39]

How Computers Make Mistakes

Among the remarkable mathematicians in the Bernoulli family, who fled to Switzerland to escape the Spanish persecution of Protestants, Jacob (1655–1705) is famous for his contributions to calculus and the field of probability and for his discovery of e, or Euler's number, a fundamental constant. His productive career included the discovery of a recipe of chaos, now named the Bernoulli map in his honour, which uses a simple rule to map one number onto the next in a daisy chain of iterated calculations. In each iteration, real numbers between 0 and 1 are multiplied repeatedly by a constant number, known by the Greek letter beta.

For Bernoulli's work, beta had the value of 2, and the rule he used to generate the map was simple: if the result of doubling exceeds the value of 1, only the remainder (what mathematicians call modulo 1) is used in the next iteration of the calculation. If you start anywhere between 0 and 1, you will bounce around in an apparently random way, eventually ending up with locations uniformly distributed between 0 and 1. This is the hallmark of chaos. Once again, this behaviour can be captured in the form of a strange attractor, and expressed in terms of orbits.

Bruce Boghosian, Hongyan Wang and Peter compared the known mathematical reality of the Bernoulli map—where the parameter beta was generalised beyond just the number 2—to what a digital computer predicted. They found that when the constant multiplier was 2, as in the original Bernoulli map (indeed any even integer), the behaviour predicted by the computer was simply wrong. The problem was not their computer or the algorithm they used but floating-point numbers.

With single precision numbers, regardless of your starting number or initial condition, you do not end up with the correct uniform distribution of mappings between 0 and 1 but always hit 0 after just 23 iterations and remain stuck there. The reason is easy to understand: in binary representation, the effect of multiplication by 2 is simply to shift the rightmost, least significant, bit one digit to the left with every iteration of the map. With double precision, it takes 52 iterations to get stuck on 0. By trying to be more precise, you simply wait a little longer for these unstable orbits to spiral into nonsense.

Let's now repeat the map with beta as 4/3, which is clearly a rational number. A little mental arithmetic (double this number, then use the remainder) reveals what mathematicians call an unstable periodic orbit, where the points 2/3 and 1/3 repeat endlessly. But when these points that constitute the periodic orbit are expressed in the binary of floating-point arithmetic, rounding becomes an issue. In binary, 2/3 is 0.10101 . . . ad infinitum, and 1/3 is 0.010101 . . . ad infinitum. Because computers cannot precisely represent these numbers, they are rounded off and, as a result, some parts of the dynamics are completely lost.

When it comes to odd values of beta, the computer's predictions work reasonably well and, according to Bruce Boghosian, many people have taken solace from this.* For the generic case of noninteger values, the computer's predictions are "not obviously wrong." Unless, of course, you take a forensic look using ensembles, which reveals errors of up to 15% compared with the exact results from continuum (nondigital) mathematics. These errors are a consequence of the degradation in the set of unstable periodic orbits that a digital computer can compute.

The bottom line is that single precision floating-point numbers are unreliable for a chaotic dynamical system of the utmost simplicity. This unsettling discovery was further investigated by Milan Kloewer at the University of Oxford with Peter, following discussions between Peter and Kloewer's colleague Tim Palmer, whom we encountered earlier when discussing ensembles in weather forecasting. Kloewer found that ensembles are able to capture more features of these chaotic systems, notably the larger orbits that not only take longer but also end up being more dominant in simulations.

The new team used double precision floating-point numbers and a method of sampling the dynamics (to use double precision would take a much bigger (exascale) computer to carry out an exact calculation equivalent to that performed by Peter's team). The periodic orbits remained very badly degraded, which surprised them. Yes, it was possible to substantially reduce the errors in the noninteger case of beta by using double precision. Nonetheless, when beta is even, the errors cannot be fixed by any improvement in the precision of the floating-point numbers, providing it is finite (which it must be to run on any digital computer). In these cases, the computer's predictions remain utterly wrong.†

Peter and his colleagues found that the error in the predictions of the Bernoulli map depends on the size of the largest periodic orbit

---

* Bruce Boghosian, interview with Roger Highfield, September 11, 2020.

† The problem can be corrected by the use of stochastic rounding or logarithmic fixed points instead of floating-point numbers.

that can be computed. This insight was followed up in joint work with Milan Kloewer on another simple dynamical system (also created by Lorenz, known as Lorenz 96). Unlike the Bernoulli map, however, there is no analytical solution, and the only way to explore the behaviour of Lorenz 96 is to experiment—and that means seeing how it behaves as it is fed with floating-point numbers of varying precision.

Despite this limitation, this study revealed another problem: for all real-world mathematical descriptions of chaotic dynamical systems, such as those that arise in molecular dynamics, as well as weather and climate forecasting, these orbits are so enormous they are not computable. However, the numerical work at Oxford did show that, at least for Lorenz 96, as the number of variables in the problem increases to many hundreds, the statistical behaviour of the system becomes very similar for all the floating-point numbers—its numerical behaviour at half precision agrees well with that at single and double precision, increasing confidence that it is correct while showing that throwing more bits at such problems—by using higher precision numbers—has no effect on the results, as anticipated by Peter with Bruce Boghosian and Hongyan Wang.

The many questions raised by this work underlines the need for more research to address how numerical results converge on the correct solutions as the degree of precision of the floating-point numbers is increased. In the wake of their 2019 publication, Peter and his colleagues discovered that Tim Sauer, a professor of mathematics at George Mason University in Virginia, along with others, had arrived at related conclusions almost two decades earlier, reporting that "extremely small noise inputs result in errors in simulation statistics that are several orders of magnitude larger."[40] He also found, in computer simulations of deterministic dynamical systems, "floating-point rounding errors and other truncation errors contaminate the simulation results."[41]

Despite Peter's work and that of Sauer and his colleagues, the overwhelming majority of those who use computers assume that floating point arithmetic is the lifeblood of simulations of any natural system that can be described in mathematical terms, from windy weather to turbulent blood flows. Little work is done to check the results of the computer simulations against the answers given by continuum mathematics. For simple systems that behave in a linear way, that is a reasonable assumption. But for chaotic systems, which are ubiquitous, our faith in computers to generate reliable results is unfounded. Whatever rational numbers we feed into calculations of chaotic systems, let alone the subset that we use for computer simulations (invariably floating point numbers), they will end up converging on unstable periodic orbits that eternally repeat, which is the opposite of chaos. Only initial conditions based on uncomputable irrational numbers are capable of producing trajectories which exhibit chaotic motion, so that they endlessly hop around these periodic orbits, never settling on one and never repeating. The latter chaotic trajectories generated by the irrational numbers—which are infinitely more common than rational numbers (the latter are said to be of 'measure zero', the former of 'measure one')—lie beyond the reach of digital computers.

Curing Digital Pathologies

The implications of chaos and digital pathology are a worry when it comes to Virtual You. Increases in the precision of floating-point numbers may not always be enough to deal with chaos, which is found in cardiac cells, axons, pacemaker neurons and molecular dynamics. Indeed, some have speculated that chaos abounds in biology because it offers "healthy flexibility for the heart, brain and other parts of the body."[42] This problem is compounded by the digital pathology uncovered by studies of the Bernoulli map.

Peter has spoken about this issue at numerous meetings. The response has been mixed. There are some who place their faith in the output from a digital computer rather than exact mathematical reasoning. They sometimes add, in a hand-waving way, that the fundamental fabric of reality is discrete because it consists of atoms and molecules. In the appendix, we explore the remarkable implications for fundamental physics were this to be borne out, though so far there is no evidence that spacetime really is "grainy."

While the hand-wavers are in denial, many others would rather not know: for them, ignorance is bliss. For complicated real-world examples, there are no known "true," exact, continuum mathematical descriptions that can be used as a benchmark. Many, faced with this dilemma, prefer to take refuge in computer simulations, even if they may be flawed.

There are ways to minimise the problem of how to be both precise and accurate that are worthy of mention. One of these is to use "stochastic rounding," a means of exploiting higher precision floating-point numbers to interpolate randomly what numbers to insert in the final part of the binary numerical representation so as to prevent the single precision floats from veering off through round-off errors into junk. This approach is not widely available on current computers but may arrive in a few years.

Others have put their faith in the "ergodic theorem" (discussed in our book *The Arrow of Time*), which originated a century ago in the study of physics problems, such as the motion of billiard balls, molecules in a gas and celestial mechanics. Ergodicity expresses the idea that, for example, the molecules of gas in a tank will eventually visit all parts of the tank in a uniform and random sense; ergodic theory studies the global properties of such systems that evolve in time. With enough time, all the trajectories spread out uniformly to produce a time-independent probabilistic equilibrium state, a property known as mixing in ergodic theory, analogous to the way a drop of black ink spreads out to make water uniformly murky when it reaches equilibrium.

Many adherents of ergodicity believe that, if they run one simulation for "long enough" it will explore all the possibilities that one could extract from an ensemble. Here they are thinking of the behaviour of isolated dynamical systems that obey Newton's equations of motion (think of billiard balls bouncing around on a hole-free and frictionless pool table), which Poincaré himself showed return to their initial state almost exactly after a finite but extremely long period of time, so they cycle forever.

But when it comes to chaotic systems that are mixing, such as the dispersing ink droplet, the probability distribution reaches an equilibrium state that no longer changes with time, so the time evolution is irreversible (which is why we discussed it in *The Arrow of Time*). Moreover, the longer one integrates a chaotic system on a computer, the less accurate the calculations become because of the butterfly effect: rounding errors can knock the system's orbit onto a very different one. Chaotic dynamics once again means that using floating-point numbers to calculate long trajectories produces unrepresentative behaviour.

Although we know very little about how the precision of floating-point (or other discrete) number schemes affects the ability of computers to converge on the right answers, one could adopt a pragmatic position that might prove acceptable in some engineering and other "real-world" applications. This argument maintains that we have substantial uncertainty both in the numerical solutions of the equations we are interested in, and in whether the equations themselves correctly describe the physical system we seek to model. We also have large uncertainties in the parameters we enter into these equations, from the use of finite—and in some cases, such as in weather and climate modelling, very crude—discretisation methods, as well as the fluctuations that arise in experimental measurements. In light of all these uncertainties, some believe there is no point in even trying to perform these numerical calculations using higher precision floats. But this purportedly pragmatic position simply plays the many sources of uncertainty against one another and falls far short of the standard of rigour demanded by a professional mathematician.

For the time being, ensembles still offer the best way to handle chaos. They provide a way to understand the problems created by floating-point numbers, and the extent to which they can undermine simulations, because the pathology survives all the ensemble (and time) averaging you can throw at it, regardless of the precision of the floating-point numbers.

However, there is yet another way to make sure that our simulations are precise, one that focuses on hardware: analogue computing offers a fully reliable way forward. As we will see in Chapter Nine, the answer could come from going back to the future of computing, developing modern forms of the first, analogue, computers that manipulate light rather than dials, spokes and gears. Simulations run on analogue computers will not display the pathologies caused by discrete number systems.

More Theory, Please

Mathematics has shown how to chart change in the body, illuminate the limits of computation and reveal ways to overcome them. But there is also a very practical issue to resolve when it comes to the use of theory in biology: mathematics does not have the long-standing relation to the life sciences that it does to the physical sciences.[43] Evolutionary theory, for example, has only been under development since the start of the twentieth century, when a simple equation to show the effect of passing genes down the generations was found independently by the German doctor Wilhelm Weinberg (1862–1937) and the English mathematician G. H. Hardy (1877–1947), best known for his work with Srinivasa Ramanujan (1887–1920), the autodidact mathematical genius who died tragically young.[44]

Today, it is possible to make some limited predictions based on various assumptions, for instance, of the evolution of the gut bacterium *E. coli*[45,46] or of the strains of the virus responsible for influenza.[47] However, there is no easy way to predict an influential mutation that can cause a discontinuous change, such as the development of a new species. Even when "all historic data" have been collected and analysed, the only way this feat of prediction might eventually be realised would be in a highly probabilistic way, using ensembles with hundreds if not thousands of uncertain parameters in play. While theories such as general relativity or quantum electrodynamics are well established, there is still argument over detailed aspects of evolution.[48] In the spirit of what Virtual You hopes to achieve in terms of placing biology and clinical medicine on firmer mathematical foundations, Hardy once told Bertrand Russell: "If I could prove by logic that you would die in five minutes, I should be sorry you were going to die, but my sorrow would be very much mitigated by pleasure in the proof."[49] Chaos has undermined Hardy's vision, but, by using ensembles, at least his desired deadly prediction can be expressed in terms of probabilities.

To spur progress with Virtual You, biology needs more theory. By this we mean something quite specific: mathematical structures that can make sense of known facts; equations, formulae and relations that reveal the relationship between quantities, and parameters that correspond to things we can measure in the real world; and mathematical structures that are backed by experiments and rigorous statistical analysis of the results and that are practical to use, too, when expressed by a computer program. And by this we mean they must be able to turn data into predictions at a later time and deliver these predictions before they become a reality.

This work is in progress, but to complete that second step will require a cultural revolution in science. We would hope to see more biology students able to understand the theory of dynamical systems, more theory-led work in biological and medical journals and more attempts to turn qualitative understanding into quantitative understanding of biology. More attention needs to be given to biological theory across a range of disciplines if the many attempts at integrating computation, big data and experiment are to provide useful knowledge.

The big data generated by experiments are paramount, of that there is no doubt. However, the impact of all these data is diminished without big theory. Though blind big data approaches are faster and easier to apply, theory and mechanistic explanations are crucial if we are to figure out whether correlations in data are true so that we can convert intriguing associations into fundamental understanding.

Developing theory that captures the mathematical quintessence of biology is hard because of the extraordinary complexity of living things. Even so, many advances have been made since the work of Hodgkin and Huxley. Moreover, when used wisely, we will see in Chapter Four that AI can help develop models.[50] Now we are ready to take the next step. Based on the insights of data and experiment gathered by Bacon's bees, simulations on digital computers are the "third leg of science," and the third ingredient of Virtual You.

3

From Analogue to Digital You

"If we were to name the most powerful assumption of all, which leads one on and on to an attempt to understand life, it is that all things are made of atoms, and that everything that living things do can be understood in terms of the jiggling and wiggling of atoms."

—Richard Feynman[1]

During the past century, computing has soared in power at a rate unmatched by any other technology, providing new opportunities for science and medicine, not least the creation of digital twins. Computers are central to the third step required to build Virtual You. They are already routinely used to animate mathematical theories that can describe the workings of the human body, even down to the "jiggling and wiggling" of its component atoms, as Richard Feynman (1918–1988) famously put it.

Though Bacon's bees are the equivalent of today's scientists, the computer has profoundly changed their relationship with the ants and spiders. That tool has provided the means for ants to manipulate a glut of data, enabled spiders to explore vast cobwebs of reason, and helped the bees to blend data and reason into predictions about what could happen next within the body, making these forecasts more easily and at a far greater scale than hitherto possible.

What would Bacon have made of it all? He may well have been familiar with a "computer" since the earliest written use of the word can be found in a 1613 book by the English poet Richard Braithwaite. In *The Yong Mans Gleanings*, Braithwaite writes: "I haue read the truest computer of Times, and the best Arithmetician that euer breathed." *Com* originates from the Latin for "together" and *puter* from the Latin *putare*, meaning "to think, or tidy up, while doing accounts": in Bacon's Day, a computer was a person who did calculations.

In one sense, Bacon would not have been surprised that mathematics could be mechanised this way. He believed *ipsa scientia potestas est*, that "knowledge is power," and espoused the necessity of scientific progress. In 1625, he wrote an essay "Of Innovations" and talked about time as "the greatest innovator." With enough time after Bacon wrote these words, the power of his bees to manipulate knowledge would be amplified to an extraordinary degree by the computer.

We will see how the foundations of Virtual You date back to 1950, when a computer was first used to vivify the pioneering theory of the English biophysicists Alan Hodgkin and Andrew Huxley that we encountered in the previous chapter. A consequence of this is that, while Hodgkin and Huxley devised a model that today is seen as a central pillar of neuroscience, their work also marked the birth of computational neuroscience. By then, other scientists had started to simulate the world from the bottom up. Today, this effort to simulate atomic "jiggling and wiggling" of atoms of molecules—including the molecules of life—is called molecular dynamics, a term also used to describe how molecules interact with each other.

The field of molecular dynamics, along with all aspects of simulation, rests on many advances in computer software and hardware, along with a lineage of technology and ideas that date back millennia. For reasons that become apparent in Chapter Nine, even the first primitive computing machines are relevant to the development of Virtual You.

Antikythera

One of the earliest—and most remarkable—examples of a computer is the astounding Antikythera mechanism of ancient Greece, a two-millennia-old analogue machine that could mimic the movements of the heavens.[2] The hand-cranked geared mechanism used pointers to recapitulate the movements of the five planets the ancient Greeks could see with the naked eye (and calculated other phenomena besides, such as eclipses).* The mechanism is the earliest known example of a machine for predicting the future.

---

* Magdalini Anastasiou, interview with Roger Highfield, August 24, 2021.

FIGURE 16. A fragment of the Antikythera mechanism. (National Archaeological Museum, Athens, No. 15987; Wikimedia Commons, CC BY-SA 3.0)

This analogue computer—an ancient Greek supercomputer, if you like—was a product of the first great scientific revolution during the Hellenistic civilisation, from around 330BCE to 30BCE. Using the Antikythera mechanism, the ancient Greeks could represent the theories they had developed of how the planets moved. By the nineteenth century, the English polymath Charles Babbage (1791–1871) had mechanised calculation with his Difference Engine, where toothed wheels carried digits to solve complex problems (figure 17). For Virtual You, we need a machine that can both represent the theories of how the human body works and handle colossal quantities of data.

Simulation science—the third ingredient of Virtual You—was born in earnest during the Manhattan Project to create the first nuclear weapons, one of the most momentous and controversial

FIGURE 17. Babbage's Difference Engine No 1. (Photo by Carsten Ullrich; Wikimedia Commons, CC BY-SA 2.5)

collaborations of the twentieth century. Among the many problems that had to be solved was working through the mathematics of explosive shock waves—captured in the form of partial differential equations, including the Navier-Stokes equations of fluid dynamics we encountered earlier. The challenge was working out how these shock waves could be focused by "lenses" to crush fissile material into a critical mass, when an out-of-control chain reaction releases colossal amounts of destructive energy.

The Manhattan Project initially relied on calculators and computers that, like the Antikythera mechanism, are analogue; that is, they harness measurable physical entities, such as distance or voltage, to do computations. Slide rules are another example, along with hand-cranked calculators. However, the passage of the high-explosive shock within a nuclear warhead required legions of calculations because it depended on more complicated partial differential equations. Each calculation had to be checked and rechecked.

There had to be a less laborious way to handle all these equations, and it was the eminent and influential Hungarian mathematician, John von Neumann (1903–1957), who introduced the weaponeers to one of the first electronic digital computers, the ENIAC (electronic numerical integrator and computer), which began operation in 1945 at the University of Pennsylvania. Relatively powerful IBM punched-card machines soon followed.[3]

When studying the feasibility of a two-stage thermonuclear weapon, where an atomic—fission—bomb is used to trigger a hydrogen—fusion—bomb, the fluid dynamics became enormously more complicated. Mathematical simulations, from the heating of a thermonuclear weapon to the fluid dynamics of its explosive bloom—known as the "Super problem"—were central to the development of the world's first hydrogen bomb, Ivy Mike, which was detonated on the Eniwetok Atoll in the Pacific Marshall Islands in 1952.

While ENIAC was used for some of the early calculations, a Los Alamos version—the mathematical analyzer, numerical integrator, and computer, or MANIAC—was conceived (and given this name, ironically, because he hated such acronyms) by the Greek-American physicist Nick Metropolis (1915–1999) and used for simulated weapons experiments.[4] This pioneering use of computers paved the way for simulating all kinds of flows, not least of blood through the human heart.

The Manhattan Project also provided key insights into how to simulate events at the molecular level, thanks to a key figure who would make seminal contributions to computational science, and who helped figure out how to ignite a thermonuclear reaction:[5] Stanislaw Ulam (1909–1984), whose wife, Françoise, was one of a cadre of women "computers" at Los Alamos. He remarked that it was the largest mathematical effort ever undertaken, "vastly larger than any astronomical calculation done to that date on hand computers."[6]

Ulam would come to simulate a random process, that of a nuclear chain reaction, through what became known as the Monte Carlo method. This influential method gets its name because it relies on computer-based (pseudo) random number generators (akin to the randomness observed in the games played in the casinos of Monte Carlo) to run the simulations. This randomness enabled Ulam to find solutions more quickly.

The results are extracted by averaging the outputs of vast numbers of separate simulations, yet another example of how ensembles—in this case "ensemble averaging"—can make sense of random processes. Metropolis, working with others, would in 1953 unveil the best-known form of the Monte Carlo method, which uses repeated random sampling to, for example, study how neutrons diffuse through a fissionable material like plutonium. This proved a quick and effective way of generating the thermodynamic equilibrium state (when overall, or macroscopic, change has ceased) of an assembly of molecules in any phase of matter—solid, liquid or gas—without having to go into the details of how molecules actually move, that is, their molecular dynamics.[7] Instead, in what is now called the Metropolis Monte Carlo method, random numbers are used to make random moves of particles, atoms or molecules after randomly chosen initial conditions to work out the average, equilibrium state.

This early work and the many other innovations in computer science and computer-based modelling at Los Alamos—along with parallel nuclear weapons efforts in the former Soviet Union and the UK—laid the foundation for many techniques now used for the analysis of fluid dynamics in weather forecasting, nuclear reactor safety, drug discovery and myriad applications, including, of course, virtual human simulations.

An alternative to Ulam's Monte Carlo statistical methods emerged in the 1950s, at what is now called the Lawrence Livermore National Laboratory in California, when Berni Alder (1925–2020) and Thomas Wainwright (1927–2007) pioneered simulations of the actual movements of atoms and molecules based on Newton's laws of motion, solving Newton's equations when atoms collided. The pair used tiny time steps to follow the resulting atomic dance on the UNIVAC, the first electronic computer in use at Livermore, which could follow 100 rigid spheres, calculating roughly 100 collisions per hour. When they graduated to an IBM-704 computer, they could handle 2000 collisions per hour for a few hundred particles.[8]

Using this alternative approach, which soon came to be known as classical molecular dynamics, Alder and Wainwright showed in 1957 that systems of hard spheres undergo a phase transition from liquid

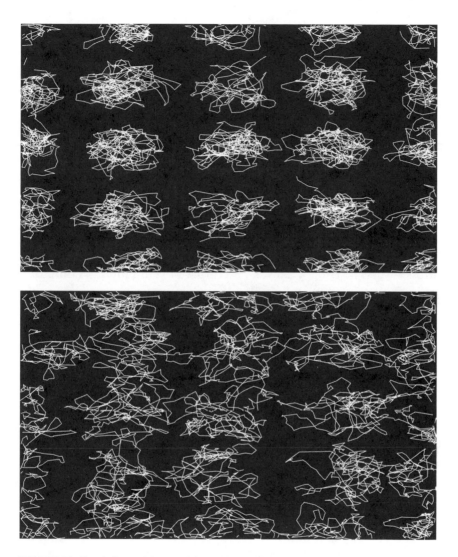

FIGURE 18. Simulating melting. Particle paths in molecular-dynamical calculations. In solid state (top), particles move around well-defined positions; in fluid (bottom), they travel from one position to another. (Shunzhou Wan, Centre for Computational Science, UCL)

to solid as they are compressed (figure 18).[9] This was an important simulation because it showed that big changes in a material, such as freezing, could be understood in terms of its component molecules, marking perhaps the first seemingly realistic simulation of matter.

Rise of the Digital Heart Cell

The Nobel Prize–winning work of Huxley and Hodgkin rested on the achievements of many others. As we saw in Chapter One, measurements of nerve impulses were made possible by Kenneth Cole using the voltage clamp technique. Hodgkin had sent the proofs of his groundbreaking papers with Huxley to Cole, who was then at the Naval Medical Research Institute in Bethesda.

While Huxley used a hand-cranked calculator, Cole decided in 1950 to run the model on the Standard Eastern Automatic Computer in Washington. This pioneering example of computational neuroscience took half an hour to calculate the 5-millisecond duration of one kind of action potential. Cole had joked that his computer model was 16 "Huxley power." A couple of years later, a computer consisting of four floor-to-ceiling racks of vacuum tubes would do the calculation in seconds.[10] This is known as the wall clock time, to express how long it actually takes to do a simulation: sometimes the wall clock time runs ahead of reality, sometimes it runs behind it and occasionally, as we discuss later, lags so far behind reality that performing the simulation is impractical.

By 1959, the year that Andrew Huxley had used the Cambridge EDSAC computer to show oscillations in the nerve model,[11] a young English scientist called Denis Noble began to apply these insights to heart cells. Noble was working at University College London under the supervision of the physiologist Otto Hutter (1924–2020), who had done important work on the pacemaker cells of the heart. Could the 1952 insights of Hodgkin and Huxley be applied to cardiac impulses?

There were many obstacles in Denis Noble's path to a virtual heart cell: the cardiac fibres he studied were tiny compared to the squid giant axon (roughly 50 μm, or 50 millionths of a metre, compared to up to 1000 μm for the squid). Nor did Noble, only an undergraduate, know anything about computers. Moreover, he had given up mathematics at the age of 16 and had no knowledge of calculus. Unsurprisingly, he had struggled to make sense of the Hodgkin-Huxley paper.

Once Noble had recorded the electrical activity of cardiac cells, he wanted to know if his mathematics could explain the way they contracted rhythmically. He signed up for a course to familiarise himself with calculus. At first, he resorted to using a hand-cranked manual calculator but realised to his horror that a 500-millisecond cardiac action potential would take him months to calculate.

To speed things up, Noble resolved to make use of an early valve computer in Bloomsbury, where University College London is based. The Ferranti Mercury was a 1-tonne 10,000 flop machine that, he recollected, had "no screen, no graphics, no windows, very little memory, and not even Fortran to help the user—the programming was done in a mixture of machine code and a primitive structure called autocode."

Time on the computer was precious, and in huge demand. "I believe I was the only biologist in the whole university to have dared to ask for time on Mercury," he recalled. Noble was rejected and told, "You don't know enough mathematics and you don't even know how to program!"

He bought a book on programming, mugged up on how to write code, and eventually returned to the custodians of the Ferranti Mercury to bid for time on the machine. "I sketched out on a piece of paper the cyclical variations in electrical potential recorded experimentally, showed them the equations I had fitted to my experimental results and said that I was hoping that they would generate what I had recorded experimentally. A single question stopped me in my tracks. 'Where, Mr Noble, is the oscillator in your equations?'"

His interrogators, only too aware of the regularity of a heartbeat, were looking for a circuit that produces a periodic, oscillating electronic signal, such as a sine or a square wave. Noble later reflected that what he should have said, based on his knowledge of the Hodgkin-Huxley model, was that "the oscillator is an inherent property of the system, not of any of the individual components, so it doesn't make any sense for the equations to include explicitly the oscillation it is seeking to explain."[12]

In the end, it did not matter this was an emergent property. Noble's sheer enthusiasm—from his emphatic hand-waving to sketching out the details on the back of an envelope—persuaded the computer's custodians to give the young biologist the graveyard shift, from 2 a.m. to 4 a.m. Now he could use the Ferranti Mercury to animate the mathematics of the various cardiac feedback loops, where the ion channels depend on the voltage, which changes as they push charged atoms in and out of cells, akin to the "Hodgkin cycle" in axons.

There were important differences, however. While a nerve has a rapidly advancing wave, or action potential, moving over thousandths of a second, the equivalent process in the heart takes much longer, around half a second. Other differences emerged in terms of the number of differential equations and ion channels (a channel can be represented by more than one differential equation).

Sodium channel activation and inactivation processes required Hodgkin and Huxley to use (at least) two coupled differential equations. They also needed one for the potassium channel and another for the axon potential, making four coupled equations in all. In his cardiac model, Noble kept their formulation of two differential equations for the sodium channel, while tailoring the constants for the heart. In light of his experiments with Hutter, he then departed radically from the equations used by Huxley and Hodgkin for the potassium channel, changing the time course by a factor of around 100, and added a coupled differential equation for the cell potential. In all, recalled Noble, "Huxley and Hodgkin had three, plus the necessary differential equation for cell potential. I had four, plus the necessary differential equation for cell potential."*

To create his simple virtual heart cell model, Noble lived a "crazy schedule" for stretches of two or even three days, starting with a night on Mercury, then a visit to the slaughterhouse at 5 a.m. to gather fresh sheep hearts for his experiments, which carried on until midnight; then, revived by strong coffee, he would start to modify the next paper tapes to be fed into Mercury by 2 a.m. He had to be on hand if things went wrong and, in the following two hours, could simulate what today takes milliseconds on a laptop.†

After a few weeks, Noble passed a key milestone. When he plotted out the results that poured out of a teleprinter, "The results did look very pretty." They revealed oscillations—his virtual heart cell was beating! After a few months, he successfully submitted two papers to the journal *Nature* to explain how rhythm was an emergent property

---

* Denis Noble, email to Roger Highfield, January 16, 2021.
† Denis Noble, interview with Roger Highfield, August 28, 2020.

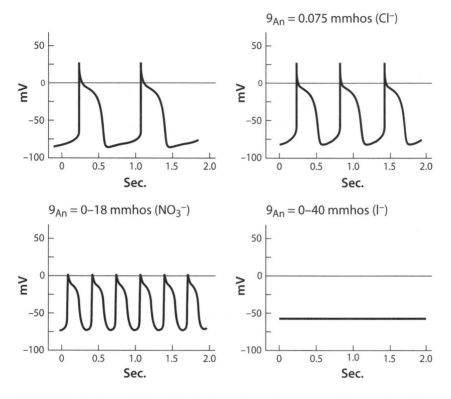

FIGURE 19. Cardiac oscillations. Adapted from "Anion Conductance of Cardiac Muscle" by Denis Noble (PhD thesis, 1961)

of components of the Purkinje fibres, the conducting fibres that tap out a heartbeat. The details appeared two years later, in the *Journal of Physiology*, in which Noble wrote: "It can be seen that the general shape of the action potential . . . corresponds very closely to that observed experimentally in Purkinje fibres."[13]

Noble, who moved to Oxford in 1963, would develop his heart cell model with Dick Tsien, Dario DiFrancesco, Don Hilgemann, and others, such as the New Zealander Peter Hunter, expanding the repertoire of differential equations to 10 variables in 1975,[14] then to 16 a decade later.[15] The first model of a ventricular cell, which incorporated a new understanding of calcium currents, emerged in 1977.[16]

By the 1980s, Noble and his colleagues had a reasonable understanding of all the activities—electrical, chemical and mechanical—

involved in the contraction of a single cardiac cell. Heart cell activity was boiled down to around 30 equations that captured the key chemical processes, notably the channels that allow electrical signals to flash in and out of cardiac cells. In this way, Noble's original model has evolved into more than 100 cardiac cell models used today, from the sinoatrial node that is the heart's pacemaker to atrial cells, and of various species too, from rabbits to guinea pigs.

Like a scalpel, these models can dissect out what is important when it comes to making a heartbeat. Failure can be as illuminating as success. Figuring out why a model does not produce the right rhythms can be as revealing as building a model that works first time. As Noble put it, "It is one of the functions of models to be wrong," since a spectacular failure can bring to light flawed assumptions and shortcomings in our understanding.[17]

Towards Virtual You

Computing has come a long way since the development of early electronic computers, such as the Colossus in Bletchley Park, which shortened World War II by cracking Nazi codes; the Ferranti Mercury used by Noble; or the Apollo Guidance Computer—a 70-pound, 16-bit computer that landed two men on the Moon.[18] Our ability to put the spark of life into mathematical equations has been transformed in recent years by the rise of high-performance computing—supercomputers. [19]

Performance has soared. While the ENIAC could manage a measly 400 flops in 1946, that figure reached 9 million flops two decades later in the CDC 6600, widely considered to be the first commercially successful supercomputer.[20] Number crunching rose to more than 300 million flops in the 1975 vintage Cray-1 (named after Seymour Cray (1925–1996), who had once worked at Control Data Corporation).[21] In 1985, the Cray-2 became the first machine to break the 1-billion flop barrier, using multiple processors.

Today, we are moving from the petaflop era to that of exascale machines, where exa- denotes three orders of magnitude better capabilities (in either performance, memory capacity or communications) compared to their petascale predecessors. These computers will be capable of at least one million million flops, that is, a one followed by 18 zeros. To match what an exascale machine can do in just one second, you would have to perform one calculation every second for 31,688,765,000 years.

This next generation of exaflop behemoths is winking into life: in Japan, Fugaku; in China, Sunway Oceanlite, Tianhe-3 and Shuguang; in the US, Aurora, Frontier and El Capitan, and in Europe, Jupiter. Never satisfied, scientists and computer engineers are already thinking about machines that are a thousand (zettascale) or a million times (yottascale) more powerful again.

The Struggle for Exascale

This leap in computer power to the exascale era has not come so easily. The earlier petascale machines had surfed on two technological waves. The first was Moore's law, named after Gordon Moore, who observed in 1965 that the number of transistors on a microprocessor doubles every 18–24 months, a trend that held up for more than three decades.[22] The second was Dennard scaling, named after Robert Dennard, which refers to the ability to increase clock speed—processing per second—while decreasing processor feature size.[23]

All good things must come to an end, however. Dennard scaling broke down around 2005. Moore's law also became bogged down. A year or two later, the consequences for high-performance computing were laid bare by the US Exascale Working Group, led by Peter Kogge of the University of Notre Dame. The group realised the projected power requirement was more than a gigawatt, drawing as much power as 110 million LEDs, 1.3 million horses, a million or so homes or sufficient power, if you believed the movie *Back to the Future*, to send a DeLorean car through time.

To sidestep heat would mean going broad (more processors) rather than deep (faster processors) to reach higher performance. As a result, exascale machines are becoming "fatter rather than faster."[24] A supercomputer is not simply a large or fast computer: it works in a quite different way, using parallel processing instead of the serial step-by-step processing in old-fashioned desktop computers. So rather than think that a supercomputer is to a computer what a Formula One car is to a VW Golf, it is more like what a fleet of VW Golfs is to a single car.

In this way, cracks in Moore's and Dennard's scaling meant that the most obvious way to increase capacity was to share the load among vast numbers of cores, or processing units. But that led to a headache with "concurrency," coordinating the number crunching across a billion computing entities, and doing it nimbly, given that each one is computing at a billion cycles per second. Given the shortcomings of traditional coding, which was not easy to scale, the challenge looked daunting.

The group's dry, downbeat conclusion was that there were "very specific major consensus challenges for which there is no obvious technological bridge with development as usual."[25] Some of their pessimism has been borne out: the decade or so pause that came as regular as clockwork between the attainment of gigaflops, megaflops, teraflops and petaflops has extended to achieve exascale.

A host of developments helped break the exascale barrier. Some advances in hardware were driven by the competitive realm of consumer electronics, according to Rick Stevens, associate laboratory director of Argonne National Laboratory, near Chicago, where the Aurora exascale machine has been developed with Intel and Cray.* The features on chips have now shrunk to a few nanometres—billionths of a metre—so their circuits can operate faster and more efficiently (by comparison, a hydrogen atom is 0.1 nanometres wide). Artificial intelligence promises yet more efficient chip design: within an area of tens to hundreds of millimetres square, chip designers have to squeeze in many thousands of components, such as logic, memory and processing units, along with kilometres of ultrathin wire.[26]

As a consequence of using more efficient chips, Aurora's power consumption has plummeted from an estimated 1 gigawatt of energy to around 60 MW, though still enough to supply 60,000 households.† To cool its hot heart, water is taken from a nearby shipping canal through 3-foot-wide pipes. Overall, the cooling infrastructure is three times the size of the computer itself.

---

* Rick Stevens, interview with Roger Highfield, August 13, 2020.
† Rick Stevens, interview with Roger Highfield, September 28, 2020.

Running a fast and efficient simulation on an exascale machine demands "codesign," that is, tailoring the architecture of the machine to the software, and vice versa. For example, the most relevant data for a program should be stored near the processing logic to overcome the "von Neumann bottleneck," the limited data transfer rate between processor and memory. That task has become more complex as the architecture of supercomputers has evolved from lots of identical processors to a structure based on nodes. Like Russian dolls, nodes contain a number of processors, within which there are numerous "cores." In turn, these cores contain central processing units (CPUs). In the fastest machines available today, the nodes also contain accelerators—mainly GPUs (graphics processing units) that originated in the gaming industry—that can perform lots of arithmetic operations much faster than regular cores and expend far less energy per flop. This heterogeneous architecture means that important hardware includes the interconnects, the ultrafast links between cores, accelerators and nodes in these behemoths, along with the software protocols and standards such as MPI, or the message passing interface, to move massive amounts of data about.

While parallel processing allows a computational task to be divided into smaller jobs across multiple processors, these jobs still have to advance one step at a time. The result is that, while we would like to understand what a virus does in a cell over a timescale of minutes or hours, these molecular dynamics simulations are typically limited to a few nanoseconds, albeit if left to run for many days or months one can reach microseconds and even milliseconds. While some predict an exascale era of "computational microscopes" that could—subject to chaos, of course—make simulations of billions of atoms tractable to understand the details of how molecules interact, we need to find ways to speed up these molecular dynamics simulations, which are on time steps of the order of a femtosecond—quadrillionths (or millionths of a billionth) of a second.[27]

To address the "parallel-in-space" bottleneck, some computer scientists are pursuing what are called "parallel-in-time" methods, which date back to work in the 1960s by Jürg Nievergelt at the University of Illinois Urbana-Champaign.[28] While conventional parallel computing consists of many operations done simultaneously, a parallel-in-time approach tackles several time steps simultaneously. Parts of the solution later in time can be computed in an iterative way before the solution earlier in time is known. One example is the "shooting method," where a coarse guesstimate that is quick to compute is followed by iterative corrections, rather like a gunner closes in on a target with each shot, based on where each successive shell lands. This underpins one of the best-known parallel-in-time approaches, called the parareal algorithm.[29]

Extending parallel computing to time as well as space has helped tame turbulence, which was famously called the most important unsolved problem in classical physics by Richard Feynman.[30] Peter and his colleagues have shown how to translate the strange attractor we encountered in the last chapter into the unstable chaos of fluid turbulence using a parallel-in-time method that does away with sequential integrations over many time steps, and without the errors that had first thrown Lorenz.[31]

But turbulence poses particular problems, even for parallel-in-time methods. If the energy in a flow is high, and the fluid is not viscous enough to dissipate the energy, the movement goes from orderly (laminar flow) to chaotic (turbulent flow), creating eddies that constantly break into smaller whirls and swirls and so on. Ideally, one would want to simulate down to the smallest levels of turbulent motion, at the so-called Kolmogorov scale.

However, while parallel processing seems to offer a way to do this, the complexity of these calculations roughly scales with the cube (third power) of the Reynolds number, the ratio between inertial forces to viscous forces in a fluid, and this number can be of the order of a million for a turbulent flow. This puts numerical simulation out of reach of most computers, though some petascale machines (JUGENE in Germany, Intrepid and Ranger in the US) have made progress. With the rise of exascale computing, however, the torrid details of turbulence will become even more tractable.

What to Do with an Exascale Machine?

We have already entered the exascale era. Each year, the Gordon Bell Prize is awarded for outstanding achievement in high-performance computing and in 2020 went to a team based in Berkeley, Princeton and Beijing for a molecular dynamics simulation of a 1-nanosecond-long trajectory of more than 100 million atoms on the Summit supercomputer, developed by IBM for the Oak Ridge National Laboratory in Tennessee. This simulation required exaflop speed, that is, 1 quintillion operations per second.[32]

The performance milestone was also reported in 2020 by the distributed computing project Folding@home, in which users download a program that can commandeer available processing power on their PCs to tackle biomedical problems.[33] When all of its 700,000 participants were online, the project had the combined capacity to perform more than 1.5 quintillion operations per second.[34] In May 2022, the Frontier supercomputer at the Oak Ridge National Laboratory—cooled by 6000 gallons of water every minute—clocked at 1.1 exaflops, with a theoretical peak performance of 2 exaflops, as measured by the benchmark test used to assess the top 500 machines on the planet.[35]

The possibilities that beckon in this new era of computation are limited only by the imagination. Exascale machines will be able to run simulations of particle interactions for experiments at CERN's Large Hadron Collider (LHC), the 27-kilometre-circumference smasher near Geneva, where evidence of the Higgs boson was presented in 2012. When it comes to the deluge of data from the Square Kilometre Array (SKA), the world's biggest radio telescope—consisting of thousands of dishes and up to a million low-frequency antennae in Australia and South Africa—two supercomputers will be required. Coming at a rate 100,000 times faster than the projected global average broadband speed in 2022, the digital downpour that is expected from the SKA will be the largest big data project in the known universe.

The heavens will open with the help of exascale machines. For example, the Aurora exascale effort is developing "Computing the Sky at Extreme Scales," or ExaSky—so that cosmologists can create "virtual universes," charting the cartography of the cosmos at the extreme fidelities demanded by future sky surveys at different wavelengths. When a telescope goes into operation, exascale machines will generate model universes with different parameters, allowing researchers to determine which models fit observations most accurately.

These titans of the information age can help us predict and understand future calamities. The Energy Exascale Earth System model should more accurately reproduce real-world observations and satellite data, helping determine where sea-level rise or storm surges might do the most damage.[36] The code has also been used to study the effects of a limited nuclear war. A team at Lawrence Livermore National Laboratory in California simulated an exchange of 100 15-kiloton nuclear weapons (each roughly equivalent to the "Little Boy" bomb that was dropped on Hiroshima) between India and Pakistan, using the Weather Research and Forecasting model to simulate black carbon emissions from the nuclear fires, then fed the results into the Energy Exascale Earth System model to show that the smoke from 100 simultaneous firestorms would block sunlight for about four years.[37]

Exascale machines, though power hungry, will help many of us to live more sustainably. More than 50% of the global population lives in urban areas and that number is rising. As cities become smarter, and more real-time data are gathered, there will be new opportunities to use exascale computers to chart the impact of infrastructure, technology, weather patterns and more. As just one example, simulations are being used to plan water management in the lower Rio Grande Valley at the southernmost tip of Texas and northern Tamaulipas, Mexico.[38]

These computational workhorses will also be used in the quest for fusion power, which offers the prospect of a clean, carbon-free source of energy on a huge scale. Using exascale machines, one can build simulations to devise ways to prevent major disruption of fusion reactions, such as the destructive release of the fiery plasma from inside its magnetically confined trap. One form of AI—deep reinforcement learning, which we explore in the next chapter—has shown promise when trained to do the challenging job of trapping the hot, inherently unstable plasma with magnetic fields.[39] In this way, fusion technologies can be tested virtually before being used.[40] From the physics perspective, fusion presents one of the most challenging simulation problems, with a need to couple around nine separate levels of description. Once again, however, this challenge is dwarfed by what is required to build Virtual You.

Can We Trust Computers?

As computers become ever more powerful, can we rely on them to make science more objective? Leaving aside the fundamental limits to computing explored in the last chapter, computers are only as smart as the people who use them, the people who write their algorithms, the people who supply their data and the people who curate all those data and algorithms. This is critical because computers are at the heart of how modern science is done and are also capable of making predictions for which no experimental data are available.[41,42]

There are many documented examples of reproducibility issues in medical science, along with psychological science and cognitive neuroscience. Though the worst fears around the "reproducibility crisis" are likely exaggerated, the consequences can be profound. As one example, the infamous MMR studies by British activist Andrew Wakefield have paved the way for a surge in antivaccination views, which are predicted to rise, according to one analysis of the views of nearly 100 million people on Facebook.[43]

Reproducibility in computer simulations would seem straightforward, even trivial, to the uninitiated: enter the same data into the same program on the same computer architecture and you should get the same results. In practice, however, there are many barriers: if you approached a team to check a simulation done a decade ago, could they persuade their old code to run?[44] Overall, it can be challenging, if not impossible, to test the claims and arguments made in published work based on computer simulations done years ago without access to the original code and data, let alone the machines the software ran on.

There are other hurdles to overcome. Some researchers are reluctant to share their source code, for instance, for commercial and licensing reasons, or because of dependencies on other software, or because the version they used in their paper had been superseded by another or had been lost due to lack of backup. Many important details of the implementation and design of a simulation never make it into published papers. Frequently, the person who developed the code has moved on, the code turns out to depend on bespoke or ancient hardware, there is inadequate documentation and/or the code developers say they are too busy to help. There are some high-profile examples of these issues, from the disclosure of climate codes and data to delays in sharing codes for COVID-19 pandemic modelling. If the public are to have confidence in Virtual You, then transparency, openness and the timely release of code and data are critical.

The extent to which we can trust Virtual You rests on what is collectively known by the acronym VVUQ. First, validation, which is confirmation that the results are in agreement with experiment, the litmus test for whether a simulation is credible ("solving the equations right"). Second, verification, which means ensuring that software does what it is supposed to do ("solving the right equations"). Third, uncertainty quantification: tracing the source of errors. Achieving sufficient reliability and robustness is one of the key challenges for digital twins highlighted at a recent meeting by Karen E. Willcox, director of the Oden Institute for Computational Engineering and Sciences at the University of Texas, Austin.*

In short, we need to broaden the Royal Society's famous scientific motto, *nullius in verba*. Not only should we "take nobody's word for it," we should also extend that healthy scepticism to *in silico* science: take no machine's word for it either. However, when computer-based predictions do manage to pass muster, in terms of VVUQ, they become "actionable"—these simulations are trustworthy within understandable limits when it comes to making decisions. To address this, Peter led a European consortium, VECMA, which developed tools to build trust in simulations.[45,46]

Putting Computer Models through Their Paces

As one example of their work, conducted as part of a Royal Society initiative, Peter and colleagues found uncertainty in the predictions arising from the models developed at Imperial College London by the Medical Research Council's Centre for Global Infectious Disease

---

* Karen Willcox, "Predictive Digital Twins: From Physics-Based Modeling to Scientific Machine Learning," CIS Digital Twin Days, November 15, 2021,| Lausanne, Switzerland. https://www.youtube.com/watch?v=ZuSx0pYAZ_I.

Analysis. Imperial's models were hugely influential, having been used to help weigh up the efficacy of lockdowns and other stringent measures adopted in the UK to deal with the COVID-19 pandemic.

To study a virtual pandemic, the CovidSim simulation tiles the country with a network of cells. These are primed with all sorts of data—high-resolution population density data, notably age, household structure and size, along with information on schools, universities and workplaces. Random number generators are used to capture the vagaries of real life, such as, for example, how getting closer than 2 metres to another person increases the probability of getting the disease.

In all, the Imperial CovidSim model has 940 parameters, though the vast majority—up to 800—describe human population and demography, household size distributions, hospital demand and so on. At the core of the model are just 6 or 7 parameters.[47] By plugging in the latest information, the Imperial team could predict the future course of the pandemic based on current understanding and, importantly, test various measures to contain the spread of the virus, from school closures to face masks. The model proved hugely influential.

To study CovidSim, Peter worked with colleagues in University College London, Centrum Wiskunde & Informatica in the Netherlands, Brunel University London, Poznań Supercomputing and Networking Center in Poland and the University of Amsterdam.[48] They investigated how uncertainties in parameters affected the predictions, running the CovidSim model thousands of times in ensemble mode within the Eagle supercomputer in Poznań. With 940 parameters, the VECMA team faced what is called the "curse of dimensionality"—there are so many permutations that even a supercomputer has neither the time nor the power to explore them all. For uncertainty quantification, Peter's team focused on 60 parameters of most interest and tweaked them, showing that 19 dominated predictions and, moreover, uncertainties were amplified by 300% in the model. The distribution of deaths predicted by the model was not a normal distribution (a so-called Gaussian or bell-shaped curve) but lopsided with a "fat tail," so expected outcomes from the model were nowhere near predictions, based on mean values of the input parameters.

Their final verdict was that, although CovidSim is a sophisticated tool, it can be improved by adding realistic randomness and the use of ensembles, where they run the model many times—with different parameters—and then weigh up the probability of certain scenarios occurring, as is done in weather forecasting. They concluded that simulations provide an imperfect picture of the pandemic. Once again, to trust your predictions you must couch them in terms of probabilities.*

The Practical Limits of Computing

There are some problems that, even with a powerful exascale machine, and even with trustworthy software and unimpeachable people working on it, there is simply not enough time to succeed. An algorithm for finding the largest number in a list of N numbers, for instance, has a running time that is proportional to N. An algorithm that, say, calculates the flying distances between N airports on a map has a running time proportional to N to the power of 2, or N squared, because for every airport it has to work out the distance to each of the others.

Theoretical computer scientists categorise algorithms according to their execution times, which they measure in terms of the number of data, N, that the algorithms manipulate. Algorithms whose running times are proportional to N raised to a power are called "polynomial," or P. If an algorithm whose execution time is proportional to N takes a second to perform a computation involving 100 elements, so N = 100, an algorithm whose execution time is proportional to N cubed takes almost three hours.

If the answer to a task can be checked quickly, then it is in class NP. But an algorithm whose execution time is proportional to 2 to the power of N, that is, exponential in N, would take 300 quintillion years. When N appears in the exponent, and no clever algorithm can reduce the execution to a time polynomial in N, this is expressed by what is referred to as a problem being "NP-hard." The classic exam-

---

* Tim Palmer, *The Primacy of Doubt* (Oxford University Press, 2022).

ple is the deceptively simple-sounding travelling salesman problem, which asks: "Given a list of cities and the known distances between each pair of cities, what is the shortest possible route that visits each city exactly once and returns to the origin city?"

Exascale You

Simulations of cells, the body, infections and more on a variety of computers is explored in detail in later chapters, but preparations are already well under way for the exascale era. Researchers at the Forschungszentrum Jülich in Germany, the powerful Fugaku supercomputer in Japan and the KTH Royal Institute of Technology in Stockholm, Sweden, are creating exascale simulations of brain-scale networks that can run on LUMI in Kajaani, Finland, a European "pre-exascale" machine. By 2014, they had developed software capable of simulating about 1% of the neurons in the human brain with all their connections, according to Markus Diesmann, director at the Jülich Institute of Neuroscience and Medicine. These simulations include the basal ganglia circuits, which sit at the base of the brain and, among other things, coordinate movement. One objective of this work is to investigate the role of these circuits in Parkinson's disease, which causes stiffness and trembling.

But there was a catch, in that the simulation took over the main memory of a petascale supercomputer, which typically consists of tens of thousands of computing nodes. To create the virtual connections between virtual neurons, this team of Japanese and European researchers first sent electrical signals to all nodes, each of which then checked which of all these signals were relevant, creating their particular part of the network. But this could not be scaled up to the exascale: the memory demand on each node of checking the relevance of each electric pulse efficiently would, at the scale of the human brain, require the memory for each processor to be 100 times larger than in today's supercomputers.

To make better use of the parallelism in exascale machines, a new algorithm was developed to first work out which nodes exchange neuronal activity and then send a node only the information it needed.

Not only did this prepare these simulations for the exascale, it also boosted the speed of simulations on current supercomputers.[49] In the case of one simulation on the JUQUEEN supercomputer in Jülich, Germany, the new algorithm cut the time to compute one second of biological time on 0.52 billion neurons connected by 5.8 trillion synapses down from almost half an hour to a little over five minutes.

As another example of the many exascale initiatives under way, Argonne National Laboratory is working with the University of Chicago, Harvard University, Princeton University and Google to use vast volumes of high-resolution imagery obtained from brain tissue samples to understand the larger structure of the brain, and the ways each minute brain cell—or neuron—connects with others to form the brain's cognitive pathways, first in a "normal" brain and then in one that has been raddled by disease or aging. High-performance machines are necessary because to compare millions of neurons and the connecting synapses among them in two human brains is a monumental challenge involving exabytes of data, let alone to study many brains at different stages of development.

Cancer is another target of exascale machines. To create a "cancer patient digital twin" for personalised treatment, the Frederick National Laboratory in Maryland has spearheaded a substantial collaboration between the National Cancer Institute and the US Department of Energy.[50,51] Eric Stahlberg, director of Frederick's Biomedical Informatics and Data Science Directorate, has been discussing modelling cancer at the exascale since 2014 and now believes "we are at the cusp of transformative research."

The collaboration shows the value of "team science" that straddles disciplines and organisations. One project, under Matthew McCoy at Georgetown University, aims to simulate one million pancreatic cancer patient digital twins to weigh up drug sensitivity and resistance. Other elements of this digital twin effort are focused on breast cancer, non-small cell lung cancer and the deadly spread of the skin cancer melanoma. As one example, "My Virtual Cancer," led by Leili Shahriyari of the University of Massachusetts Amherst, aims to model the evolution of breast cancer.

The hope is that one day cancer patient digital twins will use real-time data to adjust treatment, monitor response and track lifestyle

modifications. Consider a patient with the blood cancer acute my-
eloid leukaemia, who has been treated with stem cells from a donor
to rebuild their immune system. In the wake of a relapse, a digital
twin could help show the impact of various treatment scenarios, with
different combinations of drugs at different doses, to help discuss
the options that suit the patient. Over the longer term, legions of
digital twins of leukaemia patients could help policymakers figure
out where best to invest resources (figure 20).

Europe is also backing various major virtual human initiatives,
such as EuroHPC, which is funding the European Processor Initiative to build its own processors and accelerators, with extreme scale
supercomputing in mind. The French company SiPearl is designing
a high-performance and low-power microprocessor suitable for the
exascale, for example. There are relevant strands in Europe's vast
Horizon 2020 funding programme, including Peter's CompBioMed2
centre. Allied to this are other projects, such as DeepHealth, which
is using deep learning, computer vision and other methods to more
efficiently diagnose, monitor and treat disease, and Exa Mode, which
makes sense of big and diverse healthcare data.

Note that both the European and American initiatives are driven,
in part, by efforts to make sense of the tsunami of patient data with

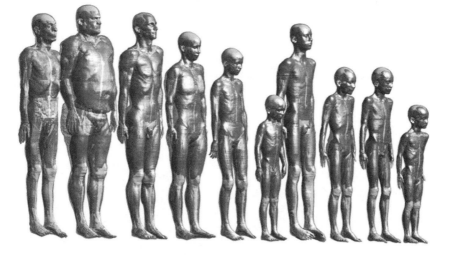

FIGURE 20. Virtual population models are a set of detailed high-resolution anatomical models
created from magnetic resonance image data of volunteers. (IT'IS Foundation)

the help of deep learning, a form of artificial intelligence (AI). The growth of computer power, which has increased at a rate unmatched by any other form of human technology, has been accompanied by rise of AI to the point where today it challenges the idea that human cognitive power is unique. AI now carries a huge burden of expectation that it will provide revolutionary insights, not least into the workings of the human body. Could AI do away with the second step required to create Virtual You and simply turn your medical data into your digital twin? The use of AI does indeed mark the fourth step towards Virtual You, not because of big data per se but what we call "Big AI."

4

Big AI

The Analytical Engine has no pretensions whatever to originate any-thing. It can do whatever we know how to order it to perform. It can follow analysis; but it has no power of anticipating any analytical relations or truths. Its province is to assist us to making available what we are already acquainted with.

—Ada Lovelace*

The field of artificial intelligence, which comes in many forms, has seen many false dawns over the past half century. The latest resurgence is particularly compelling, having coincided with a blaze in computing power and the era of Big Data. According to IBM, 2.5 exabytes—2.5 million billion bytes—of data were generated every day by 2012.[1] By 2025 the global "data-sphere" is expected to swell to 175 zettabytes (trillion gigabytes, or 1,000,000,000,000,000,000,000 bytes). By one estimate, around 90% of all the data now available to science has been created in the past two years—that assessment was made by IBM some time ago and, no doubt, the pace is accelerating.[2]

Like forty-niners panning for nuggets in the Gold Rush, many scientists today use AI to sift Big Data for meaningful patterns. Within this rising tide of measurements, facts, binary, digits, units, bytes and bits, some claim there lie gleaming nuggets, the answers to basic questions in the biosciences, social sciences or medicine, or indeed in any domain where computers wrestle with complex problems.

---

* Appendix note A to her translation of Luigi Federico Menebrea's "Sketch of the Analytical Engine Invented by Charles Babbage Esq." In *Scientific Memoirs: Selected from the Transactions of Foreign Academies of Science and Learned Societies, and from Foreign Journals* (ed. Taylor, R.), vol. 3, 696 (Richard & John E. Taylor, 1843).

Some even maintain that with the advent of Big Data, the traditional approach to science is obsolete: we have no more need for the complex dance between hypothesis, theory and experiment of Bacon's bees.[3] This simple, deterministic way of thinking argues that if we have quintillions of bytes of data about how the world has behaved in the past, we can infer how it will behave in the future. Why bother with theory at all? Why exert yourself to think about what these data really mean when a machine can furnish you with knowledge?

We are unconvinced that the scientific method is now obsolete, though we do believe that AI will indeed be crucial for the success of the virtual human project, marking the fourth ingredient of Virtual You. However, AI is most powerful when working hand in glove with mechanistic understanding based on the laws of nature, where AI hypotheses can be tested in physics-based simulations, and the results of physics-based methods are used to train AI. This is what we mean by "Big AI."

From Natural Intelligence to AI

The most promising approach to AI is inspired by natural intelligence, the human brain. In the late 1950s, while working at IBM, Arthur L. Samuel coined the term *machine learning* when referring to a program with the brain-like ability to learn from its mistakes to play draughts or checkers.[4] A more powerful approach to machine learning would arise from primitive representations of the human brain, known as neural networks, that date back to 1943, when Warren McCulloch (1898–1969) and Walter Pitts (1923–1969) of the University of Illinois suggested that brain cells—neurons—can be thought of as logical switches.[5]

Neural nets would take half a century to realise their promise. In the mid-1990s, around the time we wrote *Frontiers of Complexity*, there was a thaw in one so-called AI winter, when the field became mired, thanks to the rise of a new class of neural networks. These artificial neural networks, at their simplest, consisted of an input layer of neurons, an output layer and a hidden layer. The strengths of connections between the neurons were tweaked to enable the

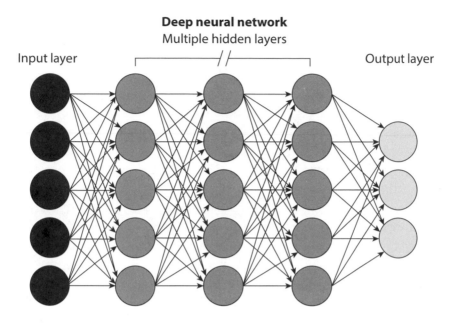

Deep neural network
Multiple hidden layers

Input layer   Output layer

FIGURE 21. A deep neural network. (IBM)

network to learn, most commonly using an algorithm called "back-propagation of errors" or backpropagation for short. With enough training data, all it took was one hidden layer to crack nonlinear problems. By exposure to pictures of felines in which they had been categorised as tigers, the network could go on a tiger hunt in a new set of images without the slightest idea of what a tiger really was.

The British cognitive psychologist Geoffrey Hinton showed that neural networks could become more powerful still by adding those extra layers of neurons—hence "deep learning," where "deep" refers to the depth of layers. These hidden layers are able to capture deeper representations of the problem, say, in categorising an image. The first layer processes a raw data input—such as pixels in an image—and passes that information to the next layer above, triggering some of those neurons, which then pass a signal to even higher layers until eventually it determines what is in the input image. Thanks to deep learning, AI is arguably enjoying its greatest success to date, whether in organising photographs, making recommendations about what to buy or what to watch, or obeying spoken commands.

Using machine learning, it has become routine to seek patterns, for instance, to separate vowel sounds or distinguish faces, excelling in all sorts of tasks, from simultaneous language translation to applications in computer vision and stock market analysis. GPT-3, created by OpenAI in San Francisco, was trained on around 200 billion words and able to write surprisingly fluent text in response to a prompt. The AI harnesses its knowledge of the statistical relationships between the words and phrases it reads to make inferences, but no one would claim that it understands their meaning. As a result, its output can sometimes be as daft as it is impressive.[6]

Deep learning has outperformed people when it comes to playing the multiplayer battle arena game Dota 2, the strategy game StarCraft II and the racing simulation Gran Turismo.[7] Games provide a limited environment, like a virtual sandbox, which machine learning can explore much more easily than the real world. It would be a mistake to think that, after learning from hundreds of thousands of games, these AIs are really smarter than people. They are simply good at these games in the same way that computers are much better than people at multiplying huge numbers. Decades of research have shown how people have a propensity to see intelligence and agency in the inanimate,[8] the curse of "apophenia" that goes back to the dawn of human consciousness, when our ancestors feared that a rustle in the undergrowth marked the presence of a malevolent forest spirit.

For this reason, machine learning can seem remarkably inventive. The deep neural network AlphaGo Zero—developed by Demis Hassabis and his colleagues at DeepMind in the UK[9]—trained in just days, without any human input or knowledge, to reach superhuman levels when playing games like chess, shogi and Go.[10] The program could not explain the strategies it developed (though, to be fair, perhaps that is the machine intelligence equivalent of what human players know as "intuition"), but, strikingly, some were unconventional and new.

There are other hints of artificial creativity. So called GANs (generative adversarial networks) are well known for their ability to create "deep fake" images, where an "artist" network creates images and a "critic" network figures out if they look real.[11] Using stacked GANs, which break down a creative task into subproblems with progressive goals, it is possible to create high-quality images from text alone, generating reasonably convincing images of birds and flowers that are entirely synthetic.[12]

Mathematics itself is being assisted by machine learning, which offers a way to tackle problems that have not been solved by other means. Recently, machine learning spotted mathematical connections that humans had missed in the study of symmetries, and helped untangle some of the mathematics of knots. In another study by DeepMind, deep learning automated the discovery of efficient algorithms for the multiplication of matrices (arrays of numbers).[13]

Another use of AI arises when mathematics becomes intractable, as Poincaré discovered even when applying Newton's laws to as few as three bodies. To make progress, rigorous and elegant theories have had to give way to ad hoc approximations. Here machine learning can help applied mathematicians out of the impasse so that, for example, it is now possible to simulate the behaviour of hundreds of millions of atoms by approximating them with an AI "surrogate," where a machine learning algorithm is trained with the results of a few well-chosen simulations that have been carried out in detail.[14] If trained properly—an important caveat—this "black box" surrogate can behave like the real thing. As one example, used to understand liquid water and ice at the atomic and molecular level, machine learning first learns quantum mechanical interactions between atoms, so that it can act as a surrogate to make speedy inferences about the energy and forces for a system of atoms, bypassing computationally expensive quantum mechanical calculations.[15]

When it comes to medicine, there is huge interest in using AI to help diagnose disease from carefully curated data. Early efforts to use computer-aided detection of breast cancer showed no extra benefit.[16] In 2020, however, there was a glimpse that AI could outperform radiologists in a study by Google Health, DeepMind, Imperial College London, the National Health Service and Northwestern University in the US.[17] The next year, AI showed promise when trained by a team based in Boston on histology slides of tumours with known origins to help find the ultimate source of "cancers of unknown primary."[18]

But AI has had its fair share of disappointments too. After studying more than 300 COVID-19 machine learning models described in scientific papers in 2020, a University of Cambridge team concluded that none of them were suitable for detecting or diagnosing COVID-19 from standard medical imaging, due to biases, methodological flaws, lack of reproducibility and "Frankenstein data sets," assembled from other data sets and rebranded, so algorithms end up being trained and tested on identical or overlapping sets of data.[19] AI methods have since made headway when it comes to diagnosing COVID-19 from chest scans.[20]

We return to the reasons why AI can run into problems later in the chapter, though it is already clear that successfully using machine learning on real-world data gathered by different doctors in different hospitals using different technologies is a much greater challenge than using neat, carefully curated training data.[21]

Solving the Grand Challenge

Perhaps the most spectacular recent use of AI relevant to Virtual You came when tackling the protein folding problem outlined in Chapter One. This grand challenge meant finding a way to turn one kind of data—the amino acid sequence of a protein—into another—a three-dimensional protein shape, which is crucial for understanding its role in a cell or body. The challenge was completed by the AlphaFold AI in 2020, according to the organisers of the biennial Critical Assessment of Protein Structure Prediction, or CASP, set up to catalyse state-of-the-art research in protein structure prediction.

AlphaFold was developed by DeepMind, a UK-based company that had already shown the power and potential of AI when it came to tackling constrained but difficult problems, such as winning games of chess and Go. By constrained, we mean that these games have limited rules, board size and pieces. Even so, like the protein folding problem, they are computationally hard because there are too many possible moves for a computer to explore all of them in a reasonable time. When it comes to protein folding, for example, a "reasonable time" means more quickly than the time it takes to isolate crystals of a protein (rarely straightforward), analyse them with X-rays or cryo-EM—cryo-electron microscopy, a form of electron microscopy with near-atomic resolution—and work out the structure (figure 22).

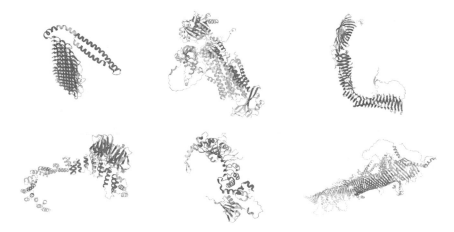

FIGURE 22. Proteins come in many shapes, which are critical for the way they work. (DeepMind)

To apply AI to protein folding, John Jumper and his colleagues at DeepMind first had to couch the problem in mathematical terms. AlphaFold looks at the known structures of sequences of amino acid residues within the proteins in the Protein Data Bank database that are related in evolutionary terms to the one of interest and then learns how to predict the structure of the target sequence from a kind of "spatial graph" that shows the relationship—the proximity—between the amino acid residues in three dimensions. From this, it can predict the position and orientation of each residue and their side chains.

The DeepMind team trained AlphaFold using sequences of amino acids and the resulting protein shapes they had adopted, stored in publicly available data on around 170,000 protein structures determined by standard experimental methods, using X-ray crystallography and nuclear magnetic resonance (NMR) spectroscopy. One caveat is that there is an assumption that all these protein structures are correct, when there are concerns that there is too much reliance today on automated methods to analyse X-ray data:[22] computers still cannot compete with chemists when it comes to analysis, though these methods are improving. Another assumption is that AlphaFold is trained with a lot of structures in which the proteins are bound to other molecules in complexes. And, of course, it should not be forgotten that the shape adopted by proteins when in the form of a cold crystal is a long way from the flexing configurations adopted in the warm confines of a cell.

AlphaFold was not programmed to predict the relationship between amino acids in the final structure, up to 2000 amino acids, but to figure out what amino acids lie near each other. A deep learning neural network, AlphaFold has many layers, and its interconnections change as its training data about protein structures are fed to the bottom layer—the input layer—and pass through the succeeding layers. As it learns during its training, the weights and thresholds of connections alter. Surprisingly, AlphaFold did not even know the amino acids exist in a chain—it just learned to place them this way. To do this, it optimised a mind-boggling 100 million parameters.

That's a lot of parameters. If you regard AI as glorified curve fitting, that is, constructing a curve, or mathematical function, that matches a series of data points, you can fit almost anything with that number. To put this 100 million figure in context, the American computer pioneer John von Neumann famously quipped: "With four parameters I can fit an elephant, and with five I can make him wiggle his trunk." The DeepMind team seemed to have enough parameters to give the pachyderm consciousness and a sense of purpose too.

DeepMind calls AlphaFold an "attention-based neural network," one able to dynamically push information around as it is trained for a few weeks on a wide range of known protein structures. As Jumper told Roger, "We see that, as the neural network starts to learn which parts of the protein are close, then it is able to establish essentially a link to pass information between different amino acid pieces. So 'attention' means that, in a sense, each part of the protein attends to, or communicates with, other parts of the protein that the network has determined might be close. So, you see it builds knowledge of the structure of this protein and then uses that knowledge to build even more knowledge about how it folds."[*]

In 2020, CASP measured the accuracy of AlphaFold's predictions in terms of the percentage of amino acids within a threshold distance from the correct position, known as the global distance test, and AlphaFold achieved a median of 92.4 (a perfect fit is 100). Using a different

---

[*] John Jumper, interview with Roger Highfield, December 9, 2020.

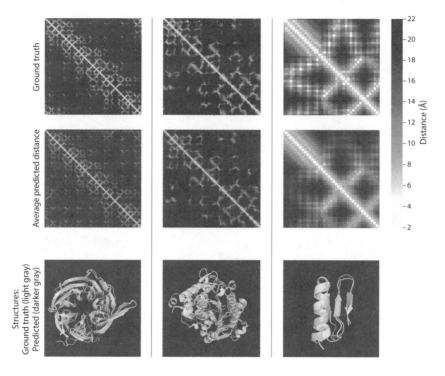

FIGURE 23. How to visualise the accuracy of AlphaFold's predictions: Here are the "distance matrices" for three proteins, where the brightness of each pixel represents the distance between the amino acids in the sequence comprising the protein—the brighter the pixel, the closer the pair. Shown in the top row are the real, experimentally determined distances and, in the bottom row, the average of AlphaFold's predicted distance distributions, which match well. (DeepMind)

metric (root mean square deviation of all amino acid residues from their experimentally reported positions), AlphaFold reported an impressively small average error of approximately 1.6 angstroms, which is comparable to the width of an atom (0.1 of a nanometre) (figure 23).

By the summer of 2021, AlphaFold had predicted the structure of nearly the entire human set of proteins, or proteome (98.5% of the full complement of human proteins) along with near complete proteomes for various other model organisms, ranging from mice and fruit flies to zebra fish and the malaria parasite.[23] These 350,000 or so protein structures—created in only 48 hours—were made publicly available from the European Molecular Biology Laboratory European Bioinformatics Institute in Hinxton, UK. Demis Hassabis of

DeepMind claimed this trove marked the biggest contribution made to date of AI to scientific knowledge as the number of predictions reached 200 million structures in July 2022, covering the known protein universe.[24] Having a reliable method to predict 3D structures of proteins will undoubtedly accelerate the speed of research in structural molecular biology.

When the paper outlining AlphaFold's methods and source code appeared, a team from the University of Washington, Seattle, published details of RoseTTAFold, another protein folding prediction program. Inspired by AlphaFold, RoseTTAFold approached the accuracy of its algorithmic muse.[25] That two different machine learning approaches have shown such success in structural biology is reassuring, but we still have to be certain we can trust all these data—they need to be validated and verified, notably when it comes to the active sites in proteins where binding takes place to other molecules, such as those used as drugs.

However, AlphaFold only provides structures for a particular conformational state, that is, a single static structure, which may fit better to the target structure. The target is usually obtained from X-ray crystallography under artificial circumstances and may well not represent biologically relevant conformations, unlike the structures measured with NMR, which produces a more dynamic and realistic ensemble of structures.

There are other caveats. AI predicted structures need to be extended to include the target molecules used for drug discovery to help create the next generation of medicines, and experimentally defined structures will still very much be needed, including notably to reveal the workings of large molecular machines using cryo-electron microscopy. Moreover, as Paul Workman of the Institute of Cancer Research in London pointed out, AlphaFold struggles with "disordered" regions, which comprise between 37% and 50% of human proteins, though it is useful to be able to identify them.[26]

There is a bigger shortcoming, however. There is still all the work yet to be done to unlock the science—the essential biology, chemistry and physics—of *how and why* proteins fold. "AlphaFold will have big impact on drug discovery," concludes Workman, "but there is no doubt that after the early stages where AlphaFold will have maximal effect, a lot remains to be done to discover and develop a drug where an accurate 3D protein structure has little to contribute."*

Why AI Can't Replace Theory

Unlike Bacon's spiders and bees, machine learning does not seek to provide explanations. That might be why the rise of AI in recent years has seen a seductive idea take hold, one that would mark a reversal of fortune for Bacon's spiders, with their obsession with the need to understand: Why bother with explanations at all? Why not do away with theory and the second step required to create Virtual You? Why not just rely on the ants and machine learning?

One reason is that the reality of machine learning is prosaic: it makes statistical inferences, a form of glorified curve fitting of known data. As a consequence, the seductive combination of machine learning and big data faces the same issues as financial services: past performance in playing the stock market is no guarantee of what will happen in the future.

There are also many examples of where deep learning has gone awry. When it comes to artificial vision, for example, machine learning can come up with bizarre classifications. Distortions that are too subtle for humans to spot[27] can fool a machine into mistaking a panda for a gibbon.[28] Abstract images can be seen as familiar objects by a deep neural net. A few well-placed stickers can make an AI vehicle misread a stop sign.[29] If these algorithms are not trained on carefully curated data, they can become maladapted, frequently embedding in-built bias of which their creators are quite unaware. At a Science Museum event that Roger helped put together, the philanthropist Bill Gates and will.i.am, an advocate for education, discussed how racism is endemic "in our system" along with gender bias, and could find its way into artificial intelligence without better curated training data, education for all and greater diversity of coders.[30]

There is also a price to pay in using deep neural networks. For every link between a pair of neurons in a neural network, one more

---

* Paul Workman, email to Roger Highfield, March 14, 2022.

connection "weight" is introduced. These parameters proliferate as roughly the square of this number, so that a large network may have tens of thousands to many millions of these weights. Success depends on optimising this vast number of adjustable parameters during training on which most computing cycles are spent.

It is often possible to tailor the network for the data with carefully selected algorithms and a well-behaved data set—one that does not contain too many discontinuities or outliers, such as a low-probability occurrence that is a big deviation from the norm, for instance, a "black swan" event in a stock market. But finding an "optimal" network means running vast numbers of these candidate networks against all manner of ways of dividing up your data set, typically by training the networks on one portion of data and then validating them on another portion they have not encountered during training. This is a little more than a trial-and-error process, one that demands a lot of computer power.

Machine learning runs the risk of what is called overfitting.[31] By this we mean that a deep neural network works well for the data used to train it but is unable to make a prediction beyond what it already knows—rather like a glove that is a perfect fit for an outstretched hand but not flexible enough to include a salute, clenched fist or Churchillian "V" sign.

This is why machine learning is so seductive and can be problematic in equal measure: with enough parameters, a neural network can always produce a snug and beautiful fit for very complex data. Hundreds of millions of parameters were used by AlphaFold to predict how the linear strings of amino acids in proteins fold up into complex three-dimensional crystal structures, but, as we mentioned, it did not do so well on the structure in solution (hence the X-ray–derived model is overfitted). Machine learning methods can easily become unreliable if extrapolating to the new and to the unexpected, which is the true form of prediction, the kind that all scientists really crave. And the reason the world often behaves in novel, unexpected ways is, of course, that it is highly nonlinear.

Recently, it has become common for researchers to use data from their simulations of a chaotic system to train machine learning to predict the behaviour of complex systems, such as turbulent flows, molecular dynamics and other manifestations of chaos.[32] Indeed, some (though not Peter and his colleagues, who bumped into the limits of floating-point numbers we encountered earlier) believe that machine learning can predict the future evolution of chaotic systems to eight "Lyapunov times," which is, roughly speaking, eight times further ahead than previous methods allowed.[33,34]

Yet many machine learning applications rely on rounded numbers (single precision and increasingly even half precision floating-point numbers) to save time and dissipate less energy. The underlying assumption is that the digital data generated by these kinds of simulations are reliable, but, given chaos, this is unlikely.

Many machine learning applications pay no attention to the way data used for training are distributed statistically. When they do focus on this important detail, many also assume the statistical distributions of chaotic data will be bell-shaped, or Gaussian, named after the great German mathematician, Carl Friedrich Gauss (1777–1855). When you make a series of measurements—heights of people, blood pressure, or indeed errors in measurements—you find that data are spread out, often around a central value with no particular bias to higher or lower values. The resulting bell curve is called a normal distribution because it is so common.

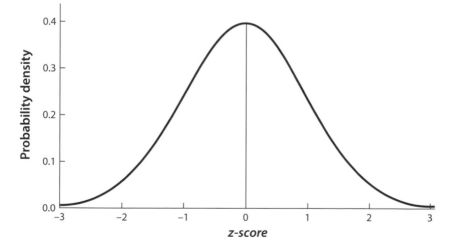

FIGURE 24. A normal distribution. Modified from "The Standard Normal Distribution: Examples, Explanations, Uses" by Pritha Bhandari (*Scribbr* 2005).

While this "normal distribution" plays an important role across science and wider human affairs, it is far from universal. Indeed, it fails to describe most nonlinear phenomena where complexity holds sway because, rather than resting on randomness, they are typically shaped by feedback loops, interactions and correlations. These complex systems—and there are few more complex than the human body—are strongly correlated, so they do not obey Gaussian statistics. To spot such phenomena, far more data need to be acquired and analysed.

Other examples of non-Gaussian statistics occur in turbulence, where the flows and vortices in a fluid extend over vastly bigger length scales than molecules, or the distant tug of gravity in the turbulent flows of matter that shape galaxies and the universe.[35] In molecular dynamics, when theory is used to predict the movements of individual molecules, the interactions between molecules extend well beyond hard-sphere-like direct collisions when electrostatic forces are in play. Other examples can be found in epidemiology and climate science,[36] such as the great currents that convey heat around the oceans. In such systems, supposedly rare "black swan" events are much more common than when viewed through the rose-tinted lens of the normal distribution.[37]

Without knowledge of the way data are distributed, machine learning can easily create serious errors. By relying on Gaussian statistics and bell curves, these difficulties can be compounded. Using imperfect data and flawed statistics to train a machine learning AI how to predict the behaviour of a system that is exquisitely sensitive to the flap of a butterfly is liable, on occasion, to produce artificial stupidity.[38,39] The good news is that, as mentioned before, computer and computational scientists are working on ways to reduce the uncertainty in such predictions, by working with ensembles of neural networks to check if they perform in a statistically reliable manner.[40]

Most machine learning approaches make another assumption that is rarely discussed. They assume a smooth and continuous curve can be drawn between a series of data points and measurements. To grasp the gist of what is going on inside a neural net, conjure up a rolling vista of hills and valleys, where the height of the landscape is equivalent to the size of the error between a prediction and previous data. In a typical application, the learning algorithm searches for the lowest feature (the lowest error) on that landscape, or search space.

A simple approach would be to crawl around in search of the lowest spot. That is fine for a gentle landscape of rolling hills. But when it comes to a jagged mountainous landscape, can you convince yourself that one dip among a landscape of rocky serrations is really the lowest spot of all, what scientists call a local minimum? This sounds a simple problem, but when it comes to real-world complexity, such as making a drug molecule dock with a target protein in the body, this is a "hard" optimisation problem since the landscape lies within a high-dimensional space. This optimisation problem is afflicted by the curse of dimensionality—one cannot get enough data and process them to make reliable predictions.

There are ways around falling into the trap of a local minimum, which we described in *Frontiers of Complexity*, among them the use of simulated annealing, genetic algorithms and the Boltzmann machine,[41] named after the great Ludwig Boltzmann we encountered earlier. Just as the random thermal motions of atoms during the annealing of a metal remove internal stresses and help a metal crystal settle into its most organised atomic arrangement, so simulated random noise generated in a computer can be made to shake the neural network out of a local dip and guide it toward the deepest valley on the error landscape.

But the real world is not always smooth and continuous. Imagine you are a golfer using a scattering of altitude data to find the 18 holes in the 300 hectares of the St. Andrews golf course in Scotland (figure 25). By their very nature, being small, holes are going to be invisible to any algorithm that joins the dots unless it has a lot of data, essentially down to a level smaller than a hole. Indeed, once these machine learners hit the superbly flat greens, their "gradient descent" algorithms would fail to work at all, leaving one concluding that the hole could be anywhere within that domain (in quantum computing, which we discuss in Chapter Eight, these are referred to as "barren plateaus"). Once again, we have to resort to the 1:1 maps of Borges to find these holes. In an analogous way, curve fitting is fine for simple systems but can be confounded by complexity and nonlinearity, which abounds in the body, from blood flow and fluid dynamics to the

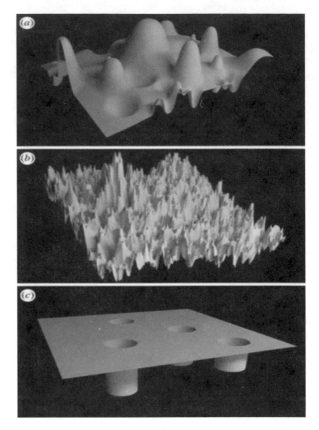

FIGURE 25.
Plots that illustrate the way in which properties of a system (the "landscape") can defeat machine learning approaches. (a) For simple situations where the landscape is relatively smooth, machine learning methods can be expected to do a good job of predicting behaviour. Examples (b) and (c) show when machine learning is likely to be challenged or to fail. Adapted from "Big Data Need Big Theory Too" by Peter V. Coveney, Edward R. Dougherty, and Roger R. Highfield (*Phil. Trans. R. Soc. A* 374, 2016).

electrical activity in the heart, the response of the immune system to invasion by a pathogen and the dynamics of molecules.

There are also practical issues: as the world faces a climate emergency, machine learning algorithms consume significant amounts of energy.[42] There is also the growing realisation that the energy demand of training and running deep learning is becoming so great that we need to find new ways to increase performance without causing the carbon emissions of computing to go through the roof.[43] One promising way around this is to no longer rely on specialist microchips but instead use analogue processing that is found in any physical system, such as manipulating light with a crystal, as part of a deep neural network. A team at Cornell University in Ithaca, New York, has shown that so-called deep physical neural networks (PNNs) can harness mechanical, optical and electrical systems to perform tasks such as image classification and vowel recognition. Physical processes are trained directly, unlike the traditional method, in which mathematical functions are trained first and then a physical process is designed to execute them. In theory, PNNs offer the way to make machine learning orders of magnitude faster and more energy efficient than conventional hardware.[44]

However, perhaps the most troubling issue of all is that we don't yet know how to build a successful neural net, in terms of the ideal recipe of layers of neurons, numbers of neurons, interconnections, network architecture and so on. The creation of a neural net remains a dark art: the only way to perfect machine learning is through trial and error. Until we understand exactly what is going on inside these black boxes, it seems unlikely that the current generation of "narrow AI," notably that based on digital machines, will easily evolve into the general AI that populates movies such as *Ex Machina* or *Terminator*.

We are indeed in the era of Big Data, but, as we pointed out in Chapter One, it takes Colossal Data to describe the human body. Biological systems are so complicated that merely selecting a particular set of data could affect the ability of machine learning to make sense of what those data describe. In a real-world system, for instance, in a living cell that is a cauldron of activity of 42 million protein molecules, can we be confident that we have captured the "right" data? Random data dredging for correlations in complex systems like the human body is doomed to fail where one has no idea which variables are important.

There is an ever-present problem of distinguishing correlation from causation. You might think, as a frivolous example, that there is no correlation between shark attacks and ice cream sales until you realise that there is another factor that provides the link: warmer temperatures encourage more people to buy ice cream as well as take a refreshing dip in the sea. Thus, this "lurking variable"—temperature—is really the cause of the apparent correlation between ice cream and injury. On top of that, too much information tends to behave like very little information. In fact, too much data can be worse than no data at all.[45] Using results from various areas of mathematics—ergodic theory, Ramsey theory and algorithmic information theory—one can rigorously show that large databases contain arbitrary numbers of correlations that increase rapidly simply with the amount, not the nature, of the data. These correlations even arise in large and randomly generated databases, which implies that most correlations are spurious.[46]

Why, given all these shortcomings, is using AI to sift Big Data so seductive? One reason is that, as we discussed earlier, biological theories have rarely reached the level of generality and power seen in physics. Much of biology and medicine relies on post hoc explanations, made after we know the answer. Once again, the reason boils down to one factor: complexity. Biological systems—notably the human body—might well be more complex than the vast structures of light and dark we call the cosmos.[47]

From Narrow to Big to General AI

Artificial intelligence will play an important role in this effort to create digital twins of the human body to deliver a new generation of truly personalised and predictive medicine. AI, notably machine learning, is already being used with some success in medicine in a variety of contexts, albeit there are plainly limits to what machine learning can do. AI surrogates will also be crucial for Virtual You. And, as we pointed out earlier, a blend of AI and mechanistic understanding—Big AI—will prove much more powerful, where an iterative cycle enables AI hypotheses to be tested in physics-based simulations and the results of physics-based methods are used to train AI.[48] In this way, the impact of AI could be transformed, not least in the virtual human project.

Big AI, or physics-informed machine learning, already marks an important extension of conventional AI, and has been used widely, from materials science to plasma physics and digital twins.[49] In fluid dynamics, for example, so-called hybrid modelling—combining physics-based and data-driven modelling—has shown advantages over using pure physics-based or machine learning models.[50] In the area of drug design, the approach has been used, for instance, in predicting antimicrobial resistance,[51] classifying the shape of enzymes[52] and predicting the way proteins bind to candidate drugs with chemical accuracy.[53]

In the United States, an AI-driven drug development platform called ATOM is one of the Cancer Moonshot task forces set up by Joe Biden while vice president.* Backed by the pharma company GSK with Department of Energy laboratories, the University of California, San Francisco, and the Frederick National Laboratory for Cancer Research, ATOM aims to integrate high-performance computing with advanced machine learning algorithms to rapidly make predictions on conceivably billions of drug candidates, which could substantially cut the time and expense of making and testing them in the laboratory, according to Eric Stahlberg, director of Biomedical Informatics and Data Science at Frederick.

Stahlberg became an evangelist for the use of high-performance computing after a family member was diagnosed with cancer. He was struck by how there was an enduring gap between the promise of advanced research and the reality of hospital care.† AI can help bridge this gap, he believes. To predict and optimise drug candidates' efficacy for a given patient, for example, scientists from Argonne National Laboratory will train different machine learning algorithms with gene expression data and model at the molecular level the interaction of the drug with cancer cells. Rick Stevens of Argonne, whom we encountered earlier, hopes that ATOM will "transform drug discovery in a big way."

In recent work with an international team of colleagues, including Stevens, we have presented a novel *in silico* method for drug design by using theory-based methods along with machine learning to make the former nimbler and the latter smarter.[54] One of the key ingredients to meld AI and physics-based modelling within a complex Big AI workflow was "middleware," software developed by Shantenu Jha at Oak Ridge National Laboratory and Rutgers University in the US.

We have used this hybrid approach for the analyses of several million compounds, and have applied it for repurposing drugs by finding those that bind to the SARS-CoV-2 main protease, the enzyme that enables the virus to reproduce, with thus far encouraging results.[55] Our colleague Rick Stevens has led an American team that has used

---

* https://atomscience.org/.
† Eric Stahlberg, interview with Peter Coveney and Roger Highfield, January 14, 2022.

a similar Big AI approach, harnessing the US supercomputer infrastructure to sift more than six million molecules to find a chemical called MCULE-5948770040, a promising inhibitor of the protease.[56]

While this approach relies on iterations between physics-based methods and machine learning to hone predictions, another kind of Big AI can be found in the guise of what is called a "physics-informed neural network," or PINN, a deep learning algorithm developed by George Karniadakis and colleagues of Brown University in Rhode Island.[57] This algorithm is both trained and required to satisfy a set of underlying physical constraints, such as the laws of motion, symmetry, energy conservation and thermodynamics. PINNs can answer questions when data are scarce and infer unknown parameters. In short, a PINN uses theory to fill gaps in data and understanding, an approach that has been applied to fluid dynamics and to improve COVID-19 prediction models, for example.[58]

Big AI represents a milestone for machine learning that lies somewhere between the current generation of AI, which has superhuman abilities albeit for very narrow applications, and a future of general artificial intelligence, when an AI agent is able to understand or learn many intellectual tasks that a human being can. Karniadakis, with Stanford engineer Ellen Kuhl and colleagues, argues that "recent trends suggest that integrating machine learning and multiscale modelling could become key to better understand biological, biomedical, and behavioural systems" when it comes to the efforts to create digital twins.[59]

When the era of Big AI is upon us, we expect it to have an impact far beyond the virtual human, where it will help build more precise models of cells, tissues and organs. The curation of AI by theory will have huge implications across science and mark an important advance towards the goal of general artificial intelligence.

Now that we have taken the first four steps, it is time to start building a virtual human. Let's examine the extraordinary range of computer simulations, from the level of molecules to the whole body, that have already taken place in the global quest to create Virtual You.

A Simulating Life

An hour sitting with a pretty girl on a park bench passes like a minute, but a minute sitting on a hot stove seems like an hour.

—Albert Einstein[1]

Albert Einstein did more than anyone to shape our understanding of space, time, energy, matter and gravity. During his annus mirabilis of 1905, when he was just 26 years old, Einstein had an epiphany that overturned Newton's centuries-old idea that space and time are the same for everybody. By assuming that the laws of science, notably the speed of light, should appear the same to all observers, he realised that space and time—united as "space-time"—depend on your point of view. Following this idea through to its conclusion, Einstein paved the way for his theory of relativity—which he popularised with the larky quote above—and the modern understanding of gravity, not as a force but as warped space-time.[2] As one physicist put it: "Spacetime grips mass, telling it how to move . . . Mass grips spacetime, telling it how to curve."[3]

Taking a lead from Einstein, it helps to recognise that there is no privileged point of reference in human biology, and that the standpoint of the emergent whole human is just as valid as one based on DNA. From the vantage point of an anatomist, the 213 bones[4] and 640 or more muscles[5] that articulate you may seem complicated. But that degree of complexity pales if you adopt another point of view, of your body from the cellular level. You consist of 37.2 *trillion* cells— that is a million million of these basic units of biology. Nothing is in charge of them, yet—acting together—they enable you to breathe, digest, move and think. When you think of the never-ending struggle to encourage collaboration among the seven billion people on our planet, the extent of molecular cooperation within your body as you read this sentence is truly extraordinary. You're a wonder.

Now imagine your body from the viewpoint of one of the tens of millions of proteins in just one of your 37.2 trillion cells. Or, as fashionable recently, from the viewpoint of your DNA code. Which level is the right place to start to build Virtual You? As Nobel laureate Sydney Brenner once remarked: "I know one approach that will fail, which is to start with genes, make proteins from them and to try to build things bottom up."[6]

While "bottom-up" thinking usually begins with DNA, there are also "top-down" effects to take into account. As one example, the type and direction of the force on a cell alters the way its genes are used.[7] We are not robotic vessels for selfish genes but a collection of cellular colonies of thousands of genes that are in dialogue with the outside world.[8] Even so, top-down and bottom-up views should give the same answers: while evolution itself was discovered by looking at a broad sweep of life, this viewpoint also makes sense in terms of DNA mutations, a beautiful example of consilience.

There is no privileged level of causality in biological systems.[9] Many regard the cell as the basic building block of the body. This might be the right level to begin to create Virtual You but for the banal reason that we have to start somewhere.[10] All that really matters is that the narrative of the body's workings from the cellular level tally with the detailed story told at the molecular level, for instance, by DNA, along with that from the viewpoint of higher levels of organisation, from organs to the body of a single person to a population.

For the time being, let's not worry about the size of each cell, or the dizzying details of its innards. Instead, let's focus on what cells do and the way they work. We can make useful progress in creating Virtual You by taking advantage of the extraordinary insights into life that have come from molecular biology, the most spectacular recent advance of biomedicine. Computer simulations can help us make sense of this rising tide of fundamental understanding.

We could try to understand how the finely orchestrated chemical choreography within a cell responds in different circumstances. Within a cell, there are many intermediates, the molecules that couple each reaction to another in metabolism, along with proteins, enzymes (which speed up chemical reactions involved in growth), energy production and other bodily functions. When the cell encounters food, for example, an enzyme breaks it down into a molecule that can be used by the next enzyme and the next, ultimately generating energy.

But rather than just observing what happens and providing a verbal account (the traditional approach adopted by Bacon's ants in biology and medicine), we must weave threads of understanding into a mathematical form. This offers deeper insights than a simple lookup table, where you link the chemistry of a cell to whether it is being starved of oxygen, invaded by a virus or activated by a hormone. This approach can also do more than machine learning that has been trained to predict how cells behave in a limited set of circumstances. A mathematical theory is—if you get it right—able to describe known behaviours and, importantly, reveal what will happen in entirely novel circumstances.

Change can be captured in the form of ordinary differential equations (though, it should be said, there are other ways of representing change, such as stochastic differential equations, which inject randomness). To put such an ODE to work, it has to be primed with experimental data about the complex networks of chemical processes within a living cell. We need the rate coefficients, or rate parameters, which express how quickly one cellular chemical turns into another.

To describe metabolic processes facilitated by enzymes, for example, scientists use the Michaelis-Menten equation, named after German and Canadian biochemists who published it in 1913.[11] The equation comes from the solution of the ordinary differential equations, which describe the rate of an enzymatic reaction depending on the concentration of the substrate, the molecule that the enzyme turns into the end products. A central factor in this equation is the Michaelis constant, a mark of the enzyme's affinity for its substrate, which has to be measured in the laboratory.

However, there is a practical limit to how much one can investigate the detailed chemistry of a cell in the lab. The networks of biochemical reactions inside these tiny units of life can involve hundreds or thousands of rate parameters, and it would take forever to determine each one by carefully probing cellular chemistry to tease out its secrets. A pragmatic alternative is to use clever ways to estimate all these parameters from a limited number of experiments on cells. One means to do this is through Bayesian methods, named after the eighteenth-century English Presbyterian minister Thomas Bayes (1701–1761), who found a systematic way of calculating—from an assumption about the way the world works—how the likelihood of something happening changes as new facts come to light. Think of it as "practice makes perfect," starting with our best current understanding and then honing that understanding with new data from experiments and/or simulations to work out the probability that a particular set of parameters fits the data optimally.

When it comes to the baroque, interconnected networks of chemical reactions in cells, the optimum parameter set is found with the help of Monte Carlo methods of the kind pioneered by Ulam, known as the Metropolis Monte Carlo scheme, after his colleague Nick Metropolis whom we encountered in Chapter Three. These probabilistic methods proved so useful that a colleague was even moved to celebrate Metropolis in a poem that referred to the "halcyon days" when "Monte Carlo came into its own."[12]

When it comes to cellular chemistry, Bayesian and Monte Carlo methods can be blended to find the likeliest (that is, the most probable) set of parameters behind a cell's networks of linked chemical processes, reaction networks. Each parameter controls the rate of one process in the network, but, in general, there will be vast numbers of these to determine. Monte Carlo methods can find the best fit of all these parameters to experimental data, namely, the way that individual chemical species wax and wane within a cell. Overall, you can think of the Monte Carlo search for these parameters in terms of random jumps around a landscape of possibilities, or search space, to find the lowest point, which is equivalent to the best fit to the experimental data.

Once we have these rate parameter estimates, we need to take into account the law of mass action, which dates back to the work of Norwegian chemists Cato Guldberg (1836–1902) and Peter Waage (1833–1900) in 1864.* This nineteenth-century law is a common-sense statement that the rate of the chemical reaction is directly

---

* The "law" is actually an approximation, because it only applies reliably to large collections of atoms and molecules, when fluctuations in local concentrations can be ignored.

proportional to the product of the activities or concentrations of the reactants in cells: more chemicals mean more chemical reactions. Equipped with numerical values of all the required parameters, the resulting differential equations can be used to describe by how much the concentrations of these species change in every step of these cellular chemical processes.

This law, though simple, can give profound insights, for instance, into chemical signalling. In 1926, Alfred Clark (1885–1941) at University College London—famous in his day for his debunking of quack remedies and patent medicines[13]—published a study of the actions of the neurotransmitter acetylcholine and the drug atropine on the frog's heart.[14] Clark used laws of mass action—developed by the American chemist and Nobelist Irving Langmuir (1881–1957) to work out how gas molecules stick to surfaces—to infer the existence of what today we know as a receptor, a protein able to receive and transduce signals. Drug molecules combine with receptors at a rate proportional to the concentration of molecules in solution and to the number of free receptors. Today, we know that atropine blocks acetylcholine receptors, for example, and that it can be used to treat nerve agent and pesticide poisonings, among other things.

Virtual Viruses

Of the millions of different chemical reactions that take place within a cell, let's focus on simulating one chain of reactions that occurs when our bodies are invaded by what the Nobelist and English biologist Peter Medawar (1915–1987) once called "a piece of bad news wrapped in protein."

We encounter viruses all the time. By one estimate, the total mass of the nonillions of viruses on Earth is equivalent to that of 75 million blue whales.[15] Measuring, typically, around 100 billionths of a meter across, viruses only come to life, as it were, when they encounter living cells. Until then, they are no more than an inert package of genetic instructions—they can come in the form of DNA or RNA— wrapped in an overcoat of fat and protein, often with protein spikes that allow them to latch onto the host cell to get inside. Once a virus has invaded our cells, its twisted genes commandeer our own cellular biochemistry with instructions that spell out a singular objective: to multiply. Following reproduction, the virus assembles itself into a virion particle that breaks out from the cell, usually causing it to die so the virus can cause more havoc.

With the growth of the global human population driving profound changes in land use, viruses that circulate in animals are increasingly spilling over to cause serious diseases in humans[16]—that was the story of HIV, of SARS, of MERS, of Ebola and of the greatest health crisis in recent decades, caused by SARS-CoV-2. Using digital twin technology, and thanks to insights gathered over the past century into viral reproduction, we can now model these infections in a computer.

When it comes to developing antiviral drugs, for example, we need to know what a virus looks like at the atomic level; here computers also play an important role, whether in making sense of X-ray diffraction or of an extension of electron microscopy called cryo-EM, which has reached atomic resolution,[17] to reveal the three-dimensional molecular structure from two-dimensional tomographic projections of individual molecules (figure 26).[18]

Thanks to studies with these techniques, and many more, we know a huge amount about how viruses work. However, although they are simple entities, they interact with the immune system of the body in subtle and complicated ways. In the case of SARS-CoV-2, for example, the rise and fall of various mutants was monitored closely during the pandemic, but it was not possible to predict how a genetic change would alter the ability of a new strain to spread and kill. It is no coincidence that the foreword of our second book was written by Baruch Blumberg, who won the Nobel Prize for his work on a vaccine that prevented millions of deaths from hepatitis B and liver cancer. He appreciated the value to medicine of the science of complexity.

Because of the elaborate nature of the immune system, to build a virtual version we need to start with one aspect of viral infection that is well understood so experimental data can guide our understanding of cause, mechanism and effect. Though dealing with COVID-19 has given a boost to the quest to use high-performance computers to simulate infection and find antiviral drugs, one to which we have

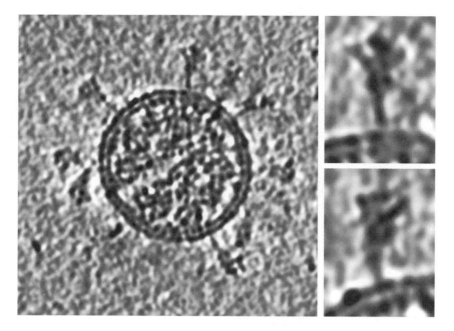

FIGURE 26. (Left) Cryo-electron microscopy of the SARS-CoV-2 virus. (Right) A close-up of the spikes it uses to invade human cells. (© 2022 MRC Laboratory of Molecular Biology)

been able to contribute, let's begin by looking at an infection that has been around for much longer, which is much better understood and where reliable computer models of infection already exist.

Virtual HIV

The human immunodeficiency virus, HIV, also spilled over from animals into people. In the case of AIDS, the pandemic originated in a virus from chimpanzees in southeastern Cameroon around a century ago.[19,20] The effects of the virus are pernicious because it targets the immune system that is supposed to protect the body.

Over many decades, researchers have studied the details of what happens when HIV attacks a type of white blood cell called a CD4 immune cell. Today, we know sufficient details to simulate key aspects of an HIV infection to test drugs, vaccines, treatments and diagnostics.

In principle, one could model infection from the "bottom up" and use molecular dynamics simulations to estimate the detailed kinetics, from the way big, deformable molecules interact,[21] along with multi-scale modelling (more later) to understand the convulsions that take place as proteins encounter each other. But it is a daunting challenge to reproduce the effects of an infection on the millions of proteins at the level of the cell, roughly 1000 times larger than its component proteins.[22,23] Moreover, these processes take place over timescales that range up to hours, an eternity by molecular standards.

Molecular dynamics can simulate the movements of vast numbers of molecules, but the time steps it must use are tiny—often no more than around 1–5 femtoseconds, where a femtosecond is one quadrillionth of a second, or a millionth of a billionth of a second (roughly speaking, a femtosecond is to 1 second what 1 second is to 30 million years). To tackle this task takes a supercomputer customised to do the job. One example is Anton, built by D. E. Shaw Research in New York, which has pushed simulations of millions of atoms into the millisecond range.[24] These simulations have to run for billions to trillions of time steps. For instance, Fred Streitz of Los Alamos and Dwight Nissley of Frederick National Laboratory used the Sierra supercomputer in Lawrence Livermore to carry out large-scale molecular dynamics simulations of cancer initiation, simulating 300 RAS proteins, which are frequently mutated in the disease. In all, they generated 100,000 microscopic simulations to model protein behaviour over merely 200 milliseconds.[25]

But, just as the map of the London Underground can get us a long way without knowing detailed geography of the city, so we can get a long way simulating an HIV infection by simply knowing the molecular players and what they do, capturing these biochemical changes in the form of differential equations. When primed with realistic rate parameters from empirical data and Bayesian methods, we can explore the impact of a virtual infection.

We can recapitulate an HIV infection with mathematics. In earlier research, for example, Peter worked with his PhD student Kashif Sadiq and a team led by Viktor Müller at the Eötvös Loránd University in Budapest to study a chemical reaction network describing the production of the HIV virus's proteins within a host cell, casting it in terms of ordinary differential equations. This mathematical model was in good agreement with many experimental observations, such as the time it takes for an HIV virion particle to mature. Armed with their model, they could also conduct a "sensitivity analysis," revealing the relative importance of various steps in the formation of the virus.[26] Such insights can help in the design of new inhibitors, drugs that can interfere with these steps to arrest the spread of the virus.

Let's focus on one particular aspect of how the body fights infections. All proteins within cells are continually being taken apart, notably in cellular subsystems (organelles known as the proteasome and the lysosome), and foreign proteins are also being degraded this way, including those from an invading virus. Ultimately, viral proteins are broken down into chunks called peptides, typically of length between 8 to 10 or so amino acid residues, which end up on the surface of an infected cell. These fragments are important because they alert the immune system when a cell has been invaded.

This is an extraordinary feat of discrimination. Your immune system has to spot a needle—peptides from invaders such as viruses and bacteria—in a haystack of vast numbers of homegrown peptides. The ability to distinguish self from nonself in this way is central to the effectiveness of the immune system. To detect infection boils down to recognising specific amino acid sequences of around 10 residues in length and working out whether they are the breakdown products of the cell, that is, self, or the remnants of an invader.

To turn this understanding into a model, we need to work out the numbers of proteins in each virus and how long it takes to crumble them into peptide pieces. This decay can be expressed as a half-life (the time span after which half of the proteins originally present in a sample have been broken down), and then the time it takes to assemble them into new viruses.[27] To produce another infectious virus, there needs to be viral RNA and viral proteins, notably three that are known to virologists by gnomic names, such as Gag, Pol and Env. We can measure the rates at which these viral components are made, together with the incorporation of these complexes into budding viruses. Reassuringly, predictions of a mathematical model of this part of the process by Peter and colleagues agree well with measurements made on the virus.[28]

In earlier work, Neil Dalchau at Microsoft Research in Cambridge, England, collaborated with Tim Elliott's team while he worked at the University of Southampton to uncover details of the kinetics, the rate of each step in the process of presenting fragments on the cell surface. First protein fragments bind to a transporter protein or chaperone, called TAP, which is in the endoplasmic reticulum and connected to a "peptide loading complex." The latter incorporates tapasin, another chaperone that promotes this marriage of molecules, along with MHC class I molecules, part of the major histocompatibility complex (MHC), which codes for protein machinery on the cell surface that can hijack the cellular waste disposal system to provide a snapshot of their contents. Overall, the protein fragments are loaded onto MHC-I molecules, then shunted through the Golgi apparatus—a cellular transport hub—to the cell's surface. For their simulations, they could represent these processes as a coupled set of ordinary differential equations.

Peter has built on all this virtual virus work to create a mathematical model of how HIV triggers a counterattack from the human immune system. Working with Neil Dalchau and their PhD student Ruth Eccleston, the team focused on the MHC genes that control how HIV peptide fragments are presented to the immune system. The molecules specified by the MHC organise the equivalent of an identity parade by holding out antigens—bits of just about everything from inside a cell—for display on its surface.

We know that if white blood cells, T cells, spot MHC molecules with "self" antigens, they ignore the cell. But if they detect foreign antigens, such as those from bacteria or viruses, they kill the cell and clone themselves to produce T cells to target other infected cells. The reality is that only a fraction of the potentially billions of peptides inside a cell are selected for presentation on its surface, and this is what the team needed to understand and thus predict how they trigger an immune response.

To model this process, the UCL-Microsoft team's first job was to translate into a mathematical form the details of the complex chemical pathway that leads to a peptide sequence on the cell surface that can be recognised by an antibody, B cell or T cell in the immune system, detailing the rate of each step. Typically, the cascading set of rate processes can be represented by many tens to hundreds of steps.

Simulations of this kind have now captured details of the development of HIV inside cells,[29,30] along with peptide optimisation, the process by which the MHC class I pathway produces a skewed sample of the incoming distribution of peptides for display on the cell surface.[31] As Neil Dalchau put it, "The peptides presented by my class I molecules are likely to be very different from the ones presented by your class I molecules. Naturally, this helps us to remain resilient to pernicious viruses, who could choose to mutate their protein sequences to avoid MHC class I binding."*

Thus, the UCL-Microsoft team could describe in quantitative terms which peptides are most likely to alert the immune system. These models predict the selection and presentation of multiple copies of viral peptides with properties that are likely to make them visible to the immune system.[32] When it comes to how T cells respond to what MHC flags up on the surface of an infected cell, once again much work has been done on the details. There are also models of how T cells respond to these viral peptides.[33]

The UCL-Microsoft team was able to wrap all this understanding into a model of how HIV triggers an immune response that could be compared with the results of experiments. Each step in the pathway, from the manufacture of an HIV protein to presentation, is represented in the form of an ordinary differential equation, based on the law of mass action. The resulting model can be used to simulate the state of a cell—infected with HIV or not—as the result of communications among proteins and small molecules that play a role in the immune system's signalling networks.

In this way, the UCL-Microsoft team could predict the presentation of HIV peptides over time and show the importance of peptides based on the breakdown of Gag proteins—which are central players in virus particle assembly, release and maturation, and function—in the establishment of a productive infection in the long-term control of HIV.

Perhaps this should not come as a surprise, given that there are around 4900 copies of the Gag protein in each virus, making Gag the

* Neil Dalchau, email to Peter Coveney, July 7, 2021.

most abundant HIV protein in infected cells. The breakdown of Gag from the virus, and from Gag newly made during cell infection, would ensure the resulting breakdown peptides are also ubiquitous, so they are likeliest to win the competition to be presented to the immune system. Simulations made with the model predict, for example, that processing of Gag can lead to a 50% probability of presentation of Gag fragments on the surface of an infected cell within three hours after infection. Reassuringly, this is also seen experimentally.[34] Critically, these simulations give far more insight than machine learning "black box" correlations, which can be useful in certain circumstances (for instance, in predicting how proteins get degraded into epitopes) but merely show that if X happens then Y is the consequence.

In subsequent work with Tim Elliott and his group, the UCL-Microsoft team was able to confirm some of the simpler predictions of the HIV model infection while opening the door to more challenging experimental investigations at the full immunopeptidome level, the entirety of the HLA peptides presented on a cell. This demands sophisticated experimental techniques at the edge of, and beyond, what is technically feasible today.[35] Even so, virtual HIV infections are becoming ever more realistic.

Virtual T Cells

The Sir William Dunn School of Pathology in Oxford has seen many extraordinary developments since it was established in 1927, from the work in 1941 on the purification and use of penicillin, which revolutionised treatment of bacterial infections, to the discovery of cephalosporin in 1961. The immune system has been a focus of interest there since James Gowans (1924–2020) showed in 1959 that lymphocytes, a key component of the immune system, recirculate from blood to lymph, and that agenda was further bolstered when mechanisms of immunological tolerance, which is critical for successful transplantation, were defined by Herman Waldmann in the 1990s.

Today, Omer Dushek's team studies T cells, which are made in the thymus gland in the chest between the lungs and are programmed to be specific for one particular foreign pathogen. As mentioned earlier, they recognise foreign proteins, so-called antigens, in the form of peptide fragments, and ignore "self" proteins so as to distinguish foe from friend. There has been a lot of bottom-up work on this complicated recognition process. Peter and his colleagues, for example, have used molecular dynamics simulations to predict how peptides bind to T cell receptors—at what is called the immunological synapse—to trigger this response.[36,37]

However, there is a limit to how easy it is to build a model from known molecular details of T cells, even though we know all (or nearly all) the proteins in T cells and that there are over 10,000 changes in phosphorylation (a protein modification) on these proteins when T cells recognise an antigen. The problem is that, to formulate this molecular model, we would need rate constants for the billions of chemical reactions within the T cells, which are currently unknown.

To sidestep being overwhelmed by details, Dushek's team instead worked from the top down, using experimental data to infer how to model T cell activation, having an effective mathematical representation of how T cells make decisions when confronted with an antigen—are they stimulated or not? In effect, by focusing on what the cells do, rather than the complex changes within them, they could use simple, high-level experimental data to constrain the possibilities offered by computer models.

At the Dunn School, Dushek's team surveyed many different bottom-up models that explain how T cells become activated and found that they can all be categorised into one of four simple ODE models.[38] But when they tested the predictions of these models, they found that none were sufficient to account for how T cells really behaved. So they systematically scanned over 26,000 different models (networks of ODEs) to identify the simplest network that could explain their experimental measurements of T cell input/output, which varied over a millionfold in the dose and the binding strength of the antigen. The best marked a simple extension of one of the original four,[39] which accurately predicted the T cell response to combinations of different antigens.

When it comes to such models, ordinary differential equations are good enough to show T cell activation and there is no need for more complex partial differential equations to chart the cellular geography of the protein machinery within them.* Dushek likens it to what a pilot needs to know: "To fly the plane, you don't need to understand all the molecular parts. If you want to 'operate' a T cell, you just need to understand the 'lever' or 'dial' that controls the T cell function. Thus, you can learn to operate (and potentially therapeutically exploit) a T cell without a detailed understanding of its very complicated insides."†

In more recent work, the Dunn School team experimentally quantified the discrimination abilities of T cells and found that they sometimes make mistakes.[40] The work may point to new ways to treat autoimmune diseases, such as arthritis, multiple sclerosis, type 2 diabetes and psoriasis, when the immune system attacks the body and the collateral damage causes disease.

The team measured exactly how tightly receptors on T cells bind to a large number of different antigens, and then measured how T cells from healthy humans responded to cells carrying varying amounts of these antigens. Their previous model could qualitatively explain their data, but this time they could quantitatively fit the model directly to their data using sequential Monte Carlo coupled to Bayesian inference. Together, their experiments and modelling showed that the T cell's receptors were better at discrimination compared to other types of receptors.

However, they also found T cells could respond even to antigens that showed only weak binding. Surprisingly, this included the normal "self" antigens on healthy cells. "Our work suggests that T cells might begin to attack healthy cells if those cells produce abnormally high numbers of antigens," said Dushek. "This contributes to a major paradigm shift in how we think about autoimmunity, because instead of focusing on defects in how T cells discriminate between antigens, it suggests that abnormally high levels of antigens made by our own cells may be responsible for the mistaken autoimmune T cell response."

There are other implications of this finding because this ability could be helpful to kill cancer cells that evolve in the body to make

---

* Omer Dushek, interview with Roger Highfield, July 27, 2021.
† Omer Dushek, email to Roger Highfield, September 5, 2021.

abnormally high levels of our antigens. Dushek told us: "We're now using this mathematical model to guide experiments to find ways to artificially improve the discrimination abilities of T cells for cancer and autoimmunity."[*]

Virtual Drugs

Drug development is a relatively mature example of virtual human research, where computers are used to tackle the nonlinear equations that abound within a cell to help develop personalised treatments. Let's say we have found a novel molecular target in your body, a protein that, when mutated, is linked with the development of disease. The hope is that we can design a drug, like machining a key to fit a lock, that works just for you by testing a virtual drug on a virtual copy of that molecular target (figure 27).

This has long been a dream of chemists, and indeed theoretical chemists were among the first who clamoured to use early digital computers. One application was in molecular dynamics, which we mentioned in Chapter Two. Another was in quantum chemistry: S. Francis Boys at the University of Cambridge in the UK used computers and quantum theory to work out the electronic structure of molecules.[41] By the time we were undergraduates at the University of Oxford, one of our lecturers was Graham Richards, who in the 1970s was using the basic computers of the day to calculate properties of molecules, as "computational chemistry" became commonplace.[42] Why not try to model how various potential drugs interact with targets in the body?[43]

Many teams employ this approach, but it would be tendentious to claim it has had a transformative impact, and Peter's team has documented the shortcomings of many of these efforts.[44] Today, the cost of developing a new drug is measured in billions of dollars[45] and takes something of the order of a decade. Each new "one-size-fits-all" drug works for around 50% of the population, roughly speaking. This may come as a surprise, but this disappointing success rate can be

* Omer Dushek, interview with Roger Highfield, July 27, 2021.

FIGURE 27. Simulation of a potential anticancer drug, binding to an enzyme linked with leukaemia. (CompBioMed and Barcelona Supercomputing Centre)

seen in studies of antidepressants,[46] neuraminidase inhibitors used against influenza[47] and the realisation that most common medications have only small to medium effects.[48]

There are many reasons why drug development is so slow and ineffective. There are an estimated 10 to the power of 60 drug-like molecules in "chemical space," that is, the theoretical space populated by all possible compounds and molecules.[49] Of this ocean of possibilities, only a few will bind to a target in the body and, even if they do, another serious hurdle is how to deal with nonlinearity to figure out if it is enough to be a useful drug.

Chaos looms large in the story of using computers to develop drugs because the "keys" and "locks" are not rigid but, as we said earlier, complex three-dimensional molecules that vibrate, rotate and deform. The tiniest change in the rotation, shape or angle of attack of a drug molecule can have vastly different effects on its ability to bind with a target, and be effective: this is the signature of chaos. Even with vast amounts of data about how drugs work, chaos means the way a drug molecule docks with a target site in the body is extraordinarily sensitive to many factors: that is why, using traditional computer methods, independent simulations of the action of an HIV drug on the same target molecule can produce different results.[50]

All is not lost, however. Peter has once again pressed Monte Carlo methods into use to quantify the amount of uncertainty produced by dynamical chaos. When it comes to molecular dynamics, such an approach has only become viable during the past decade owing to its demands: a petascale machine or bigger is necessary.[51] Just as today's weather forecasts use ensembles of predictions, seeded with a spectrum of plausible starting conditions, so ensembles of simulations can provide an accurate forecast of how well a novel molecule will bind with a target site in the body to weigh up if it could make a promising drug.

Monte Carlo Alchemy

To forecast the behaviour of novel drugs, and how well they dock with a molecular target in a person's body, Peter and his colleagues have worked with various pharma companies, such as GlaxoSmithKline, Pfizer and Janssen, using a host of supercomputers: Blue Waters at the National Center for Supercomputing Applications at the University of Illinois Urbana-Champaign; Titan and Summit at the Oak Ridge National Laboratory in Tennessee; Frontera and Longhorn at the Texas Advanced Computing Center; ARCHER, and now ARCHER2, at the University of Edinburgh, and Scafell Pike at the Hartree Centre in Daresbury, UK; in the rest of Europe, Prometheus in Krakow and Eagle in Posnań, Poland; Cartesius at SURF, a coalition of Dutch research institutions, in Amsterdam; and SuperMUC and SuperMUC-NG at the Leibniz Supercomputing Centre in Garching, Germany. The fact that he could draw on such a diverse array of powerful computers in different countries is a remarkable testament to the global nature of modern research.

Predicting the efficacy of a virtual drug in docking with a virtual protein would be a key ingredient of Virtual You and, along these lines, Peter has devised a way to rank how well a range of widely varying drug candidates stick to a section, or domain, of a protein known as the bromodomain, which is linked to cancer and inflammatory as well as epigenetic diseases. Gratifyingly, the results of these ensemble simulations were reproducible. They were delivered rapidly enough to be of practical use and were personal too, being based on the details of an individual's bromodomain.[52]

The team used their molecular dynamics protocol to model how these drug interactions occur in the watery innards of cells (hence dubbed ESMACS—enhanced sampling of molecular dynamics with approximation of continuum solvent). Once they had found a promising drug candidate for uses varying from cancer treatment to pain reduction, they could put the candidate through more stringent testing with "Monte Carlo alchemy," a reference to a method they called thermodynamic integration with enhanced sampling (TIES), where, once again, Monte Carlo–based ensemble methods are used.[53]

In this way, Peter and his team could use supercomputers to predict the transmutations from a well-understood drug molecule, via unreal intermediates, wittily named "alchemical" intermediates (since they do not exist in the real world), into an untried molecule to calculate how well the latter could work as a drug. They found that, despite the alchemy, this approach can give realistic insights into the relative ability of a drug to bind to a protein before and after a mutation.

Peter's UCL team also predicted the efficacy of drugs that interfere with key HIV-1 proteins, such as the HIV protease inhibitors. The results do indeed tally with experimental data, which is as important for drug discovery as it is essential for the scientific method.[54] That is by no means the end of the applications of computational methods to HIV, however. In related work with Peter Sloot, who led the ViroLab project, a virtual laboratory to model infectious diseases at the Institute for Advanced Study in Amsterdam, Peter and many colleagues worked on models from molecular-level interactions over nanoseconds to sociological interactions on the epidemiological level, spanning years of disease progression,[55] and tested them in six hospitals.

This kind of research, which is being conducted by many other teams, raises the prospect of being able to explore how a patient with a particular genetic makeup will respond to a particular drug. This productive form of alchemy can go beyond personalised drug discovery and can also show that common mutations in the human genetic code cause acquired resistance to anti-breast-cancer drugs, such as tamoxifen and raloxifene. The dream is that one day a doctor

FIGURE 28. This painting depicts a coronavirus just entering the lungs, surrounded by mucus secreted by respiratory cells, secreted antibodies, and several small immune system proteins. The virus is enclosed by a membrane that includes the S (spike) protein, which mediates attachment and entry into cells. (Created by David S. Goodsell, RCSB Protein Data Bank)

could take your genetic code and, within a few hours, work out the best cancer drug or combination of cancer drugs for you.

This approach has also been adapted to address antimicrobial resistance, one of the biggest threats to health today and to future medicine, by Peter's former PhD student Philip Fowler, working with Public Health England within the Nuffield Department of Medicine and the John Radcliffe Hospital at the University of Oxford.[56] Blending genome sequencing of tuberculosis microbes and alchemical free energy calculations, his team can work out how quickly resistance to a given drug could develop, generating results rapidly enough for use in the clinic.[57] The Oxford team is leading an initiative, known as the Global Pathogen Analysis System, to address COVID-19 drug resistance in a similar way.[58]

Computing Cancer

Although cancer is often portrayed as a complex disease that comes in many forms, ranging from solid tumours to blood cancers, in one sense it is easy to understand. Your body depends on exquisite coordination among a vast range and number of cells, some 37.2 trillion in all, from the muscle cells that enable you to wiggle your toes to the nerves that pass electrical impulses around your brain.[59]

Typically, though not exclusively, when there is a mutation that alters the DNA code, there is a breakdown of cooperation on the cellular level. Perhaps the mutation induces a cell to divide when it should not, or hinders cell repair mechanisms, or prevents the cell from committing suicide—a process called apoptosis—when the body tells it to self-destruct for the greater good. Cells that break ranks can, if left unchecked, proliferate to fulfil their own selfish agenda, rather than that of the body. When cooperation breaks down and the individual motives of cells begin to dominate, the result is cancer.[60]

By shrugging off the controls that constrain the rest of our body, tumour cells divide unchecked, picking up new genetic changes so they can evolve to resist drugs, or grow faster. Within the extraordinarily complex environment of the human body—which differs depending on the person—the cells in tumours face diverse selection pressures, which favour cells that accumulate mutations that make them better able to survive: these, for instance, include favouring cells which accumulate mutations to better enable their survival, cells which divide faster and cells which are less likely to perish. As a result, even a single tumour can contain utterly different genetic mutations in the cells in one region, compared with cells in another. Eventually, if the tumour has not been successfully eradicated, cells will evolve with the ability to travel elsewhere in the body. This is called metastasis and is the deadliest aspect of cancer.

Though this is a simplified picture, it underlines how cancer cells are distorted versions of normal cells, making them hard to target and destroy without causing damaging side effects, when normal cells are caught in the crossfire. Moreover, because cancer results from the selfish evolution of a person's own cells, each tumour is different.

Despite this, traditional cancer treatments are the same whoever you are. Chemotherapy, for example, relies on drugs that are toxic to all dividing cells in the body. Cancer cells divide more quickly than most cells, and are more vulnerable, but normal cells are affected too, causing side effects such as hair loss and nausea. The future of cancer treatment is to move away from these blunt one-size-fits-all treatments towards personalised medicine.

One way, used in "precision medicine," is to stratify patients with a given cancer into groups that are similar in some way. As just one example, it has long been recognised that hormonal therapy for breast cancer is most likely to be effective when the cancer contains receptors for estrogen and/or progesterone hormones. But this approach, though valuable, is backward facing, being based on earlier trials.

Supercomputers can help deliver truly predictive and personalised medicine. In 2016, Peter provided a glimpse of how to tailor cancer treatments using SuperMUC, a cluster of two linked supercomputers—just shy of 250,000 computing cores in all—capable of more than 6.8 petaflops (10 to the power of 15 floating-point operations per second) run by *Leibniz Rechenzentrum*, the Leibniz Supercomputing Centre (LRZ) near Munich.

To get access to all 250,000 or so cores of this supercomputer (it has since been replaced with a new computer with over 300,000) is unusual, but, with the help of LRZ's director, Dieter Kranzlmüller, this ambitious project was slipped in during scheduled maintenance. Peter's team had access to the entire machine for 37 hours, now known as a "block operation," the equivalent of something like a quarter of a million people beavering away on everyday personal computers for 37 hours. Even by the standards of scientists who routinely use supercomputers, this was a huge job. The giant workflow consisted of a cascading series of parallel and serial computations and was so ambitious that a methuselah of champagne was used to incentivise the team to manage the sustained and uninterrupted number crunching, which generated about 10 terabytes of data, that is, 2 to the power of 40 bytes.

During a day and a half, simulations were used to study how around 100 drugs and candidate drugs bind with protein targets, in order to rank their potency for drug development. As mentioned already, though you can think of it as designing a molecular key (a drug) to fit a protein lock in the body (such as a receptor), this analogy breaks down because of chaos, when tiny changes in the structure of these small molecules can sometimes produce huge changes in the strength of binding—what is called an "affinity cliff." Once again, a promising way forward is to use ensemble-based methods to explore the effects of riffs on shapes, energies and so on. Ensembles can reproduce real measurements of binding affinity with a high degree of accuracy and precision.[61]

Critically, they found the method is reproducible and actionable too: binding affinities could be calculated within a few hours, thereby ranking all available drugs in order of their effectiveness for a given person. In follow-up work with pharma companies, blinded predictions, which were made before experiments were complete, turned out to agree well with experimental data.

The approach has been used for breast cancer, which is linked with high levels of the hormone estrogen. In a later study, Peter's team discovered novel variants of the estrogen gene and receptor in a group of 50 Qatari female breast cancer patients. They were able to suggest which existing cancer drugs would work for these patients and which would not.[62] This kind of approach heralds the development of medicine that is customised for a particular patient and gives answers quickly enough to make a difference.[63]

Virtual Heart Drug Testing

In the wake of the work by Denis Noble on the virtual cardiac cell, remarkable advances have come in virtual trials to ensure that, once we have found a useful drug, we can be sure it will be heart safe. Licensing authorities now insist on testing for one particular form of "cardiotoxicity" by seeing if there is a longer-than-average time between the Q and T waves on an electrocardiogram, which reflects how long the heart muscle takes to recharge between beats. Early warning of possible heart problems this way is critical for the development of drugs: around 40% of drugs withdrawn from the market between 2001 and 2010 had cardiovascular safety issues. The problem is that some harmless substances, like grapefruit juice and the antibiotic moxifloxacin, also prolong the QT interval, so how can we figure out which drugs do pose a risk?

In 2008, a European Science Foundation workshop discussed further steps in developing computerised *in silico* models of the heart that enable possible drugs and therapies to be tested without risk to people. The resulting preDiCT project was officially launched on June 1, 2008, with the backing of big pharma and a mission to model, simulate and ultimately predict the impact of drugs on the heart's rhythm using computer models, and was designed to drive the development of mathematical models of individual ion channels, which control how and when cells contract.

Similar work has taken place in the United States. The comprehensive in vitro proarrhythmia assay (CiPA) was launched in 2013 by the US Food and Drug Administration to spot the risk of proarrhythmia, when a drug aggravates an existing problem with heartbeat or provokes a new arrhythmia. The test used three strands of evidence, notably computer simulations to show the effect on the electrical properties (the action potential) of a human ventricular cardiomyocyte, a heart muscle cell.

Replacing Animals with Silicon

About a decade after the launch of Europe's preDiCT project, Blanca Rodriguez's team at the University of Oxford passed an important milestone in their efforts to mimic the heart *in silico*: by then their model was accurate enough to be useful for drug companies and better than animal models. While Denis Noble's first cell model relied on a handful of differential equations, her model used 41.*

* Blanca Rodriguez, email to Roger Highfield, January 27, 2021.

   Working with Janssen Pharmaceutica in Belgium, Rodriguez's team carried out a virtual trial in which 62 drugs and reference compounds were tested at various concentrations in more than a thousand simulations of human heart cells.[64] This model did not work down to the molecular level. Nor was this a full-blown heart simulation of the kind we discuss in the next chapter. Instead, it was a large and coupled set of chemical kinetic ordinary differential equations that had been developed by Yoram Rudy's laboratory at Washington University in St. Louis with experimental data on more than 140 human hearts; it predicted the risk that drugs would cause abnormal rhythms in patients with 89% accuracy. When virtual heart predictions were compared with data from comparable animal studies, the Oxford team found the animal research was less accurate (75%). In light of this success, the team won an award from the National Centre for the 3Rs (NC3Rs)—an organisation dedicated to replacing, refining and reducing the use of animals in research and testing. Even better, now that virtual cells have been shown to be more accurate than animal experiments in predicting adverse side effects on the rhythm of the heart, virtual assays pave the way for reducing our dependence on the estimated 60,000 animals used globally each year for these studies.

   Our colleagues in the Barcelona Supercomputer Center have done similar studies to check whether potential treatments for COVID-19 could also cause heart problems. Using Barcelona's MareNostrum 4 supercomputer, Jazmin Aguado-Sierra and Mariano Vázquez looked into the antimalarial drugs chloroquine and hydroxychloroquine, as part of a broad effort to plunder arsenals of old medicines to find drugs that could be quickly repurposed to fight the pandemic. Complementing major trials, they could use the mathematical models developed by CiPA to address the cardiac safety of the drugs.

   At high heart rates and high doses, the simulations showed that chloroquine caused a "conduction left bundle branch block" when the activation of one of the heart's two main chambers, or ventricles, is delayed, so that the left ventricle contracts later than the right ventricle. As this work flagged up issues, a deeper problem emerged. A major trial concluded that hydroxychloroquine was associated with an increased length of hospital stay and increased risk of progressing to invasive mechanical ventilation or death.[65] These drugs simply did not work. Reassuringly, however, the virtual clinical trial provided remarkably similar results to recent published clinical data.[66]

We can go even further in virtual testing by screening drugs, starting from their chemical structures, for their ability to cause deadly arrhythmias. To connect the dots from atomic structures to proteins, tissues and heartbeats, manifested as gradients of electrical activity in simulated tissue, various supercomputers (Stampede2 in Austin, Comet in San Diego, Blue Waters in Urbana-Champaign and Anton 2 in Pittsburgh) were enlisted by Colleen E. Clancy and Igor Vorobyov and colleagues at the University of California, Davis, to make a direct comparison between the electrocardiograms of the simulated tissue and electrocardiograms from patients on those drugs. The aim was to identify promising lead compounds with a low risk of cardiac arrhythmia.*

They studied the hERG (human Ether-a-go-go-Related Gene) potassium channel, which mediates the electrical activity of the heart and, over billions of time steps, could simulate how a drug works. By taking this to the cell and tissue scales, they were able to model how the drugs diffuse across membranes and interact with hERG in a fibre of 165 virtual heart cells, a 5 × 5 cm virtual ventricular tissue and an *in silico* wedge of left ventricular tissue.

With multiscale computer simulation data, aided by machine learning to work out necessary and sufficient parameters that predict vulnerability to arrhythmia and by enough high-performance computing, they could use this multiscale model to distinguish drugs that prolong the QT interval, such as dofetilide, from those that are harmless, such as moxifloxacin.[67] They were also able to predict that dofetilide caused considerably larger prolongation of the QT interval in women, consistent with clinical and experimental data. We can now go beyond simulated cardiac tissue to creating a virtual organ. In Chapter Seven, we discuss how to build a virtual heart.

---

* Colleen E. Clancy, email to Roger Highfield, January 19, 2022.

Towards the Virtual Cell

We used to think there were around 200 varieties of cells in the human body. Today, we know there are many more. They come in all sorts of shapes and sizes—from flat, round red blood cells to long, thin neurons that shuffle electrical signals around the body. The cells in your skin measure around 30 millionths of a metre while, by comparison, your blood cells are much smaller, coming in at just 8 millionths of a metre across.

Even if we pick a particular kind of cell, whether a brain or muscle cell, each one is a more than a bag of complex biochemistry, having a detailed anatomy that is packed with structures. There is a central compartment called the nucleus, which contains around 1 metre of tightly coiled DNA. Dotted around the nucleus are other compartments or "organelles," such as the mitochondria, which are chemical powerhouses; the endoplasmic reticulum, mentioned earlier, where proteins are synthesized by ribosomes; the Golgi apparatus, for protein packaging and secretion, alongside structures to deal with waste, known as the peroxisome and lysosome.

Once again, there is nothing in charge of all these organelles, nor their component molecules. There is no privileged level of causation. And, as Einstein averred, the laws of nature have to tell the same story from every possible perspective. Can we create a multiscale mathematical model that links our, albeit limited, understanding of the DNA instructions that describe a cell to measurements of the details of cellular biochemistry and anatomy? There is confidence that we can, and, as we are about to learn, many teams are now able to grow virtual cells.

The Virtual Cell

A mathematician, like a painter or poet, is a maker of patterns. If his patterns are more permanent than theirs, it is because they are made with ideas.

—G. H. Hardy, *A Mathematician's Apology*[1]

By simulating the spiders' webs of chemical processes within cells, and how they vary over time, the insights of molecular biology have started to come to life within virtual cells. We have seen how, even neglecting many details, it is possible to design personalised drugs and more for Virtual You. But living things are also dynamic patterns in space. To create true virtual cells in a computer, we need to understand these patterns, find ways to capture them in the form of mathematics and figure out how to simulate them in a computer.

Around the start of the twentieth century, scientists began to uncover how chemistry can create lifelike designs. In 1896, the German chemist Raphael Liesegang (1869–1947) showed that a drop of silver nitrate seeds the formation of rings of silver dichromate in a gel containing potassium dichromate. He speculated that a similar process lay behind banded effects on the wings of butterflies and the hides of animals.

More creative chemistry was reported in 1911, when French physician Stéphane Leduc (1853–1939), in his article "The Mechanism of Life," pointed out how "living beings for the most part present a remarkable degree of symmetry,"[2] and carried out chemistry—the precipitation of inorganic salts, diffusion and osmosis—to generate a kaleidoscope of shapes in what he called "synthetic biology." Insights into the dynamic nature of chemical motifs followed in the mid-twentieth century in the Soviet Union, with the exploration of patterns in space and time—from chemical "clocks" to concentric

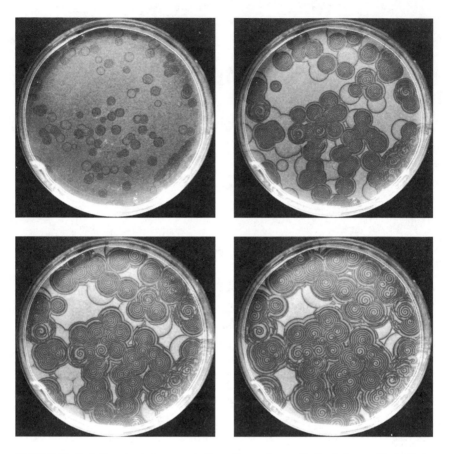

FIGURE 29. Evolution of spiral waves, similar to those observed in the Belousov-Zhabotinsky reaction, here resulting from the addition of a solution of sodium hydroxide on the top of a layer of aluminium chloride in gel. (Reproduced with permission by Manal Ammar and Mazen Al-Ghoul, Department of Chemistry, American University of Beirut)

bands and spirals—self-organising spatial and temporal structures formed by the remarkable Belousov-Zhabotinsky chemical reaction we discussed in our first book, *The Arrow of Time*.

In his 1940 book, *A Mathematician's Apology*, G. H. Hardy described a mathematician as "a maker of patterns." For mathematics to move beyond simulating networks of biochemistry within cells to capturing their vital patterns demands a shift from using ordinary differential equations to partial differential equations. These equations can deal with an apparent paradox: living things are dynamic and changing, yet chemicals can react and diffuse to create rock-steady patterns, such as an organelle, liver or body.

This paradox was solved by Alan Turing, who had helped crack the Nazi naval Enigma code during the Second World War and was also interested in the codes of life, having read D'Arcy Thompson's *On Growth and Form*,[3] which gathered together what was then known about nature's patterns—the shapes of cells, honeycombs, corals, shells, antlers, bone, plants and more. Turing would devise a mathematical model that explained how random fluctuations can seed structure, so that chemicals that are diffusing and reacting neither produce uniformity nor disorder but something in between: a pattern.

Even though Turing is best known for his work in computing, his role in biology would be cemented by a paper in the *Philosophical Transactions of the Royal Society* in 1952, published when he was 40 years old and working at the University of Manchester. Entitled "The Chemical Basis of Morphogenesis," Turing explored how a spherical (symmetrical) bundle of identical cells develop into an (asymmetrical) organism.

Turing speculated that substances called morphogens ("shape-formers") are somehow responsible for a path of development. These patterns are described by chemical rate equations, governed by the law of mass action and Fick's law of diffusion, which shows how molecules travel according to concentration gradients (the same law dictates how the aroma of garlic spreads from an open mouth to the person sitting on the other side of a table). These are combined into "reaction-diffusion equations," partial differential equations that describe how concentrations of reacting molecules vary in both space and time.

By coupling nonlinear reactions and diffusion, one can bridge the gap between the molecular world and the realm of patterns that are visible to the naked eye. In Turing's model, one chemical is known as an "activator," which is autocatalytic and so introduces positive feedback. The other is an "inhibitor," which suppresses the effects of the activator. Turing demonstrated how a reaction-diffusion system with just two chemical species could, in theory, create spots or stripes. His model could also produce stationary waves, a "dappled" pattern, like the splodges on a black-and-white cow, travelling waves of the kind that might wiggle the tail of a sperm, along with oscillations. Turing's paper marked the first time that mathematical modelling based on the laws of nature was used to illustrate how two interacting chemicals with different diffusion rates can, far from equilibrium, generate stable patterns.[4] No vitalist principle was required.

The idea that, under far from equilibrium conditions, diffusion can lead to unequal distributions of components—a stable pattern—was unexpected since forms of diffusion we encounter in everyday life usually diminish differences in concentration: fortunately, garlic breath dissipates in open air while the swirls made by hot, frothy milk added to black coffee eventually disappear to form a murky mixture, for example.

The Oxford mathematical biologist James Murray, whom we talked to for *The Arrow of Time*, came up with one imaginative, if not slightly bizarre, way to grasp Turing's patterns by considering a burning field of grass, where grasshoppers live. When his grasshoppers get too hot, they sweat profusely, so much so that they can drip on the nearby grass to stop it burning. Patterns emerge when the fire (activator) starts to spread at a given speed and, fleeing the heat, the grasshoppers move more quickly, producing sweat to damp down the flames. The result is a pattern consisting of charred and green patches of grass.[5]

Turing's work represents an important milestone, one with wider implications than simply providing insights into pattern formation in biology. He did his calculations on the Manchester Ferranti Mark 1 computer, the world's first commercially available general-purpose electronic computer, and speculated in his paper about how more detailed cases of biological pattern formation could be studied "with the aid of a digital computer."

Tragically, Turing himself was not to develop his pioneering ideas about computing the patterns of nature: two years after his paper appeared, he killed himself with a glass of dissolved cyanide salt, followed by a bite of an apple to make his last drink more palatable.[6] For the next couple of decades his work was largely ignored by chemists and biologists, who were captivated by the new insights that spun out of DNA's double helix.

Only in 1990 did hard experimental evidence emerge of a Turing pattern in chemistry.[7] Now that decades more studies have been carried out, and soaring computer power has helped tame partial differential equations, Turing's theory has offered a possible explanation for a range of markings found in the natural world, from zebras to giraffes to jaguars[8] to seashells, along with the arrangement of fingers, toes[9] and hair follicles in mice.[10] By bringing biology to life in a computer, Turing represents yet another pioneer in the effort to create Virtual You.

Project K

While Hodgkin and Huxley devised one of the most successful models of a complex biological process, their work in the 1950s was overshadowed by the molecular biology revolution that came in the wake of the 1953 publication by Crick and Watson of the double helix structure of DNA.[11] During the following decade, interest began to focus on how the proteins and molecules of life worked together within a cell. By studying simple organisms, perhaps one could grasp the detailed workings of life.

The ambition could be seen in a paper written by Crick in 1973 entitled "Project K: The Complete Solution of *E. coli*," after discussions with Sydney Brenner (1927–2019), who had been among the first to gaze upon the DNA double helix.[12] Crick and Brenner wanted to harness the new molecular understanding to attack a series of biological problems, each focused on understanding a different life-form: a virus that infects bacteria, phage lambda (Project L); the mouse (Project M); the nematode (Project N), which would lead to a Nobel Prize for Brenner; and *E. coli*, the strain K12 of the gut bacterium (Project K). When it came to the last of these, Crick talked of seeking a "complete solution" of a bacterial cell, when a living cell would be "completely" explained.[13]

In 1984, a few years after scientists began to use laborious methods to sequence individual genes, Harold Morowitz of Yale University picked an easier target for providing a complete explanation of a living cell than *E. coli*, plumping for a *Mycoplasma* bacterium ("the simplest living cells") to explore what he called "the logic of life." Despite the limitations of the computers of his day, Morowitz even thought about putting this logic to work in a computer to create a virtual cell. He talked of a computer model being "feasible . . . every experiment that can be carried out in the laboratory can also be carried out on the computer."[14] That same year, Michael Shuler and colleagues at Cornell University created a model, from the top down, based on differential equations that could reproduce the growth of a single cell of *E. coli*, and populations of bacterial cells.[15]

In the wake of this early research, efforts to create virtual cells have thrived on the rising tide of computer power, nourished by the explosion of information about the basic processes of life from molecular biology laboratories all over the world, and advances in theory, such as how the physics governing droplet formation, along with the origin of bubbles in a glass of champagne, can help us understand the basic principles that organise living cells.[16]

Rise of the Virtual Bacterium

With the rise of genome sequencing, and the glut of detailed cell data, the ability to model simpler organisms, such as the *Mycoplasmas*, became a practical reality. The vision of Morowitz received a fillip in 1995, when *Mycoplasma genitalium*, which inhabits the human genital tract, became the second free-living organism to have its genetic code read, or sequenced, in research by the American genomics pioneer Craig Venter.[17]

One attempt to turn this molecular understanding into an "E-CELL" was made by Masaru Tomita at Keio University, Fujisawa, Japan.[18] In the spring of 1996, working under Tomita's direction, students at the Laboratory for Bioinformatics at Keio charted the web of metabolic interactions based on just 120 *Mycoplasma* genes—a substantial simplification, since the real thing actually depends on 525 genes. Moreover, they also borrowed another 7 genes from the bacterium *E. coli* to persuade their virtual cell to work.*

This *Mycoplasma* model, published three years later, consisted of metabolism, gene transcription and gene translation. While they described cellular metabolism with a set of ordinary differential equations, they used object-oriented programming for transcription and translation, where the cellular data was represented as discrete objects with which the user and other objects may interact. The team

---

* Koichi Takahashi, email to Peter Coveney and Roger Highfield, October 13, 2021.

developed hundreds of rules to govern metabolic pathways, from glycolysis and gene transcription to protein synthesis. For extra realism, the enzymes and other proteins gracefully "degraded" in their model, so that they had to be constantly synthesized in order for the cell to sustain "life."

The Japanese team could experiment on their virtual cell as their "simulator engine" chugged away at around one twentieth the rate of the real thing. The state of the virtual cell at each moment was expressed in the form of concentrations of its component substances, along with values for cell volume, acidity (pH) and temperature. They could knock out a gene and study the aftermath, even "kill" their virtual cell if they disabled an essential gene, for example, for protein synthesis.

When they published their model in 1999, they had made significant progress. However, there were still many genes whose functions were not yet known. It was also difficult to build a coherent model out of ordinary differential equations. Another challenge was finding all the parameters to feed into those equations. Now led by Koichi Takahashi at the Laboratory for Biologically Inspired Computing, Osaka, the team has moved on to simulate other cells, such as red blood cells and cardiomyocytes. To avoid the chore of gleaning data from thousands of papers to model more complex cells, the team is now using AI to retrieve genomic sequences from databases, predict what parts of the sequence are translated to proteins and use existing knowledge to work out what they do and how they interact.

More recently, another virtual *Mycoplasma genitalium* has thrived in a computer thanks to the systems biologist Markus Covert at Stanford University, who himself had been inspired by the vision of Brenner, Crick and Morowitz, along with the data of Venter. While a graduate student, Covert was captivated by the first comparative genomics study in 1995, when Venter's team juxtaposed the genetic code of the bacteria *Mycoplasma pneumoniae* and *Haemophilus influenzae* to begin to work out the core set of genes required for life.[19] He can still remember the excitement of reading a newspaper report that cited a member of Venter's team, Clyde Hutchison, who talked about how a working computer model of a cell would be the ultimate test of biological understanding.

That quotation would stay with Covert, who pursued this dream. His team has used Venter's *Mycoplasma* DNA code along with data gleaned from more than 900 scientific papers, including almost 2000 experimentally observed parameters that reflected the organism's genome, transcriptome, proteome, metabolome and so on. They carved up *M. genitalium* into 28 functional processes, modelling each independently over a 1-second timescale. The whole-cell model was simulated using an algorithm comparable to those used to numerically integrate ordinary differential equations.

The change of the virtual cell's state was calculated second by second by repeatedly allocating the cell state variables among the processes, executing each of the cellular process submodels, and updating the values of the cell state variables, until a given time was reached or the cell divided. Overall, the model blended a mixture of Boolean algebra (a meld of logic and mathematics that has played a key role in the creation of computers), linear optimisation and ordinary differential equations to bring the organism "to life."

The tricky part was integrating the 28 functional processes into a unified cell model. They began with the assumption that these processes are approximately independent on short timescales—less than 1 second in this case. The cell simulations were carried out on a loop in which the models of the subprocesses were run independently at each time step, but depended on the values of variables determined by other submodels in the previous time step. In this way, they managed to model *M. genitalium* down to every one of its 525 genes—marking the first time this had been done for all of an organism's genes and every known gene function.

By 2012, the Stanford team had reported the first model of a bacterium that took into account all known gene functions and argued that the great challenge of understanding how complex phenotypes arise from individual molecules and their interactions could now be tackled with virtual cells.[20] When they unveiled their virtual bacterium, a simulation for a single cell to divide once took around 10 hours and generated half a gigabyte of data. "Believe it or not, the actual *M. genitalium* cell also takes roughly 10 hours to divide," Covert told us.*

---

* Markus Covert, interview with Peter Coveney and Roger Highfield, August 6, 2021.

The Stanford team used this virtual organism to tease out details of the cell cycle, which consists of three stages—initiation, replication and cytokinesis (cell division). In a nod to the real world, the duration of individual stages in the cell cycle varied from virtual cell to virtual cell, while the length of the overall cycle was relatively consistent. They found that cells that took longer to begin DNA replication had time to amass a pool of free nucleotides, the building blocks of DNA. As a result, replication of DNA strands from these nucleotides then passed relatively quickly. In contrast, cells that went through the initial step more quickly had no surplus of nucleotides, hindering replication.

The model provides various insights, revealing how the chromosome is bound by at least one protein within the first 6 minutes of the cell cycle. They could figure out that around 90% of the genes had been used, or expressed, within the first 143 minutes of the cell cycle, and that 30,000 interactions between proteins take place during each cycle. Within a computer, they could simulate the effects of disrupting each of the 525 genes in thousands of simulations. This revealed how 284 genes are essential to sustain *M. genitalium* growth and division and that 117 are nonessential, broadly in agreement with experiment.[21] Switching off metabolic genes caused the most disruption. When turning off the synthesis of a specific cell component, such as RNA or protein, the model predicted near-normal growth followed by decline.

They found fascinating examples of redundancy. Deletion of the lpdA gene should kill the cell according to their model. However, experiments had shown this strain stays viable, albeit growing 40% more slowly than the wild type. They deduced that another cellular subsystem was compensating. And, indeed, the team found another gene, Nox, had a similar function to lpdA. When they corrected their virtual cell by modelling this additional use of Nox, it yielded a viable simulated cell. "We were surprised by the capacity of these models to predict behaviour at the molecular scale that we subsequently verified experimentally," said Covert. This is impressive, though the team stressed that the model was only a "first draft."

One of the key insights to emerge, notably from Covert's PhD student Jayodita Sanghvi, was to go from overall virtual cell behaviours, such as the effects of gene disruption on growth rate, to predict what was happening at the level of enzymes, rather than resort to data about those enzymes from other organisms, such as *E coli*.[22] "It really was an incredible achievement that is often undersold," remarked Covert. "The earlier work was about running the simulations, and those simulations were an end in themselves. Jayodita's was the first whole-cell work in which both global and very detailed model predictions were shown to be accurate. I couldn't believe it when I first saw it—to me, these new data were as much a Holy Grail as the model itself!"*

This approach could be a boon to efforts to adapt microorganisms for novel purposes, whether to make drugs or fuel, and to inform efforts to build synthetic cells. In a tour de force of recombinant DNA technology and genetic engineering to help identify the fundamental genes for life, which took almost a decade, Craig Venter's team announced in 2016 that they had made the *M. genitalium* chromosome and variants in the laboratory and transplanted them into an empty host bacterial shell to produce a synthetic cell.[23]

This virtual bacterium, with its 525 genes, marked only the beginning. They automated the process with what they called their Gibson assembly robot (named after Venter's colleague, Dan Gibson[24]). In the long term, virtual cell models could lead to rational design of novel microorganisms. Venter, who stressed that Covert's work was "fantastic," told us that the possibilities are currently limited because there is still some way to go to fully understand gene function: even in the case of little *M. genitalium*, the role of one-third of its genes was initially unknown. However, he added that virtual cells have "tremendous utility" when it comes to laying bare these details.†

As one example, simulations have been made of the minimal synthetic bacterial cell JCVI-syn3A, a pared-down organism with only 493 genes created by the J. Craig Venter Institute (JCVI) and Synthetic Genomics, Inc. Clyde Hutchison and Zaida (Zan) Luthey-Schulten and her colleagues at the University of Illinois Urbana-Champaign built a 3D simulation of around two billion atoms that reveals the connections between metabolism, genetic information and cell growth, incorporating 148 known metabolites, 452 proteins, 503 ribosomes

---

* Markus Covert, email to Peter Coveney and Roger Highfield, September 14, 2021.
† Craig Venter, interview with Peter Coveney and Roger Highfield, December 29, 2021.

and DNA undergoing over 7000 reactions. The model relied on GPUs to simulate a 20-minute span of the cell cycle—one of the longest, most complex of its kind to date. Encouragingly, the model showed that the cell dedicated most of its energy to transporting molecules across its membrane, which is what one would expect given that it is a parasitic cell, and that fundamental "emergent behaviours are validated by several experimental results."[25]

Virtual *E. coli*

The *Mycoplasma* and minimal genome organisms are much simpler than the gut bug *E. coli*, the original target of Crick and Brenner's Project K and an organism that is much more important to science, medicine and biotechnology. This bacterium is much more complex too: *E. coli* has 4288 genes, divides every 20 to 30 minutes and has a much greater number of molecular interactions, each of which would further expand the time required to run the simulation. Others were inspired by the ambition to simulate this more complex cell, according to Covert: "Paul Allen (1953–2018), the US business magnate, researcher and philanthropist, was a big fan of this work and supported all the *E. coli* stuff financially."*

To tackle *E. coli*, Covert's team used a similar approach to *M. genitalium* and published their results in 2020.[26] They harvested more than 19,000 parameter values from the literature and databases and plugged them into interdependent mathematical equations—ordinary differential equations, stochastic variants to these, probabilistic models and so on—to simulate the interaction of cellular processes, linking gene expression, metabolism, cell replication and growth. Rather than strictly enforcing conformance with real-world data, they allowed the model to flex a little.

All proteins are regularly rebuilt in a living thing and, though far from completing Crick's Project K ("it is more like a half cell model," Covert admits), they found they could predict protein half-lives. This sounds trivial, but Covert is as sceptical of experimental

---

* Markus Covert, email to Peter Coveney and Roger Highfield, September 13, 2021.

data (different teams using different techniques can come up with different parameters) as others are of models and believes by interrogating the different sources of experimental data with their model—what he calls "deep curation"—he can come up with robust insights. As one example, protein half-lives were traditionally calculated by a rule that was established many decades ago.[27] Covert's model gave different lifetimes and, in each case, was shown to be correct by subsequent observations. These insights were important when it came to predicting the growth rate of the virtual organism. The finding was, remarked Covert, "beyond my wildest dreams."

The next step was to introduce geography into his bacterium. Enter the "molecular artist" David Goodsell, who had spent many years creating watercolours of proteins ("computational modelling was not up to creating the images I wanted") and then visualising molecules on the computer for drug design in the Molecular Graphics Laboratory at the Scripps Research Institute in La Jolla, California. As he remarked, Covert's data "showed us what the pieces are, and where they need to go." Using these data, Goodsell, Art Olson and postdoc Martina Maritan sought the molecular structures of each individual protein and put them together using cellPACK, cellPAINT and cellVIEW, software developed by Maritan's colleague Ludo Autin that exploits advances made by the gaming industry, which leads the way when it comes to state-of-the-art computer graphics.

In this way, the Scripps team developed a container—the virtual cell—and dropped in all the pieces, along with the way they interact, to create 3D pastel images of all the proteins packed into *Mycoplasma*.[28] Another program, FLEX, ensured the components fitted snugly and did not overlap. After three years of playing with the computer modelling data, they came up with an image of the crowded innards of the virtual cell that was as mind-bogglingly complex as it was beautiful, the first step towards a high-fidelity simulation.*

Covert's group has also started to simulate colonies of *E. coli*, where whole-cell models can divide, multiply and interact, whether they physically push each other around, or secrete food, waste prod-

---

* Martina Maritan, David Goodsell and Ludovic Autin, interview with Roger Highfield, May 20, 2021.

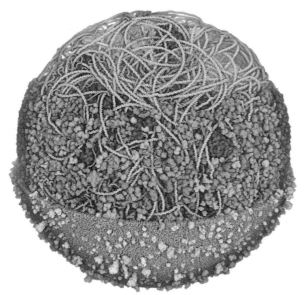

FIGURE 30. 3D model of a *Mycoplasma genitalium* cell. (Created by Martina Maritan, Ludovic Autin and David S. Goodsell, Scripps Research and RCSB Protein Data Bank)

ucts or a cleaved antibiotic. One surprising discovery emerged when they looked at how genes are used by each bacterium in the colony. They had predicted that, over a cell cycle, the amounts of protein would rise exponentially until the cell divided, when levels would drop until the rise began all over again. That common expectation proved to be the case with around one-third of the genes. But they were surprised to find most other genes were rarely made, and, when they were, the amount of protein increased sharply in single bursts, followed by a reduction by half in each generation as the cell divided, until the expression event occurred again.[*]

This puzzling behaviour can be understood when you realise that each bacterium can only hold a certain amount of protein, so it benefits the whole colony of bacteria to "spread bet" on genes, notably antibiotic resistance genes. Some bacteria bet on one bacterial invader, others on another, so the superorganism will have something at hand to respond to a threat. At the time of writing, Covert was studying what the implications of such spread bets are for operons, logical operations involving a cluster of genes.

[*] Markus Covert, email to Peter Coveney and Roger Highfield, September 14, 2021.

Getting the bigger picture is important when it comes to creating a virtual version of a bacterium. Just as weather forecasting was revolutionised by the telegraph, which "enabled news about the weather to travel faster than the weather itself," Covert believes that the next advances will come from more holistic, real-time views of the cell and its environment, research that is being carried out not just in his laboratory but, for example, in that of Luis Serrano at the Barcelona Institute for Science and Technology, who has been working on *Mycoplasma pneumoniae*.[29]

While local weather can only be predicted by having some understanding of the global context, from jet streams to warm fronts, so Covert believes predicting cell behaviour will depend on using current data. That means more than sequencing genes but developing software, AI and technologies to monitor in real time a plethora of data on the complex and teeming innards of a single cell.

VCell and ModelBricks

While bacterial cells are relatively simple, and called prokaryotes, the building blocks of plant and animal bodies, including our own, are much more complex, and known as eukaryotes. These cells contain a nucleus where their DNA resides and other organelles, the leftovers of earlier episodes in evolution of microbial mergers and acquisitions.

Take, for example, little lozenge-shaped structures in our cells called mitochondria. Not only do they look like separate entities, they even have their own DNA that is passed down from mother to child. Our cells, from muscle to brain, are driven by these descendants of bacteria that hundreds of millions of years ago traded chemical energy for a comfortable home in another cell. To model them accurately, we have to move from ordinary differential equations that deal with changes in time to partial differential equations, which can handle several independent variables. To wrangle these equations, we also need much more computational power.

In parallel with the efforts to model *Mycoplasma*, another initiative winked into life at the University of Connecticut in the mid-1990s.[30] There Les Loew convinced faculty members to share their expertise, time and equipment for live cell imaging methods at the then Center for Biomedical Imaging Technology so that different disciplines—cell biology, chemistry, optical engineering, mathematics, physics and computer engineering—could foster an open development environment to allow teams to try out modules of virtual cellular chemistry. While the *Mycoplasma* efforts came up with bespoke models of one organism, the Connecticut team wanted to create an assembly line to make all sorts of virtual cells.

Before construction of a spatially explicit partial differential equation model, a typical user of their virtual cell—VCell—starts with ordinary differential equations.[31] Within cellular compartments, the software enables reactants and products to be linked, along with the enzymes that turn the former into the latter. VCell facilitates the development of multicompartmental ordinary differential equation models, for instance, to simulate reactions in the cytoplasm as well as within the nucleus. At this point, partial differential equations can be introduced so that the concentrations within cellular compartments can evolve as functions of both space and time. To work, one needs to quantify how the molecules diffuse, and these diffusion coefficients are gleaned from experiments (using techniques such as fluorescence recovery after photobleaching (FRAP), and fluorescence correlation spectroscopy (FCS), where proteins are tagged with fluorescent markers).

The geography of a real cell is incorporated into the virtual version with the help of confocal microscopy, which has revolutionised our picture of the cell in the past two decades with its stunning three-dimensional, colourful images. These images can be used as a backbone for experimental data to build 3D spatial models of the networks of chemical reactions. The geometry is then subdivided by the user into a uniform mesh of elements—as ever, one big enough to be useful but small enough not to be too computationally intensive. Reaction, diffusion, the movement of heat or matter, transport across membranes and electrophysiology are supported by VCell. Teams can feed VCell with measurements they have made of real cellular processes, organise these data, share them and use mathematical models to simulate and analyze the biology on this modular computational framework.

With his colleague Michael Blinov, Loew took us through their model of the cell, which shows individual biochemical reactions as nodes.* Click on a node and you can see that chemical process at work. The chain reactions of chemistry are shown as networks, and these in turn can be mapped to the cellular geometry, whether confocal microscope images of slices through the cell or even idealized shapes, such as spheres and cubes. The user can specify the initial conditions of the cell and the model can also work out the concentration gradients of cellular processes, showing patterns along the lines predicted by Turing.

At the Mount Sinai School of Medicine and Columbia University, this virtual cell has been used to study how human kidney cells are regulated to produce a delicate filtration system, consisting of fingerlike projections that allow the kidney to function.[32] Other projects use the VCell to study pancreatic cells[33] and the way in which signals in the brain are transferred across synapses and neurons.

This virtual cell software is now used by scientists worldwide to study how the architecture of cells shapes and controls their response to their environment, in the way cellular chemicals react and diffuse. The model was recently updated to include cell movement too.[34] Loew told us: "Our software is designed to very generally model cell biology, accounting for diffusion with PDEs (as well as ODEs and stochastic events). This makes it valuable to a broad range of investigators—our registered user count stands at 24,000."†

Importantly, it also offers a database that can be grazed by other users to find models (or fragments of models) to build upon as they develop their own models, which they in turn can lodge in the database, typically when they publish a paper describing their cell simulations. At the time of writing, almost 1000 public models could be found in the database and about 275 have been fully curated. "As the database continues to grow, it will in a very real sense (pun intended) be the Virtual Cell," Loew told us.

Most computational models in biology are put to a single use. Blinov aims to create highly annotated and curated modules— dubbed ModelBricks—that users can mix, match and modify to build complex virtual cell models.[35] While the traditional bespoke way of

* Les Loew and Michael Blinov, interview with Roger Highfield, October 17, 2020.
† Les Loew, email to Peter Coveney and Roger Highfield, July 13, 2021.

making a virtual cell "is akin to building a brick house starting with a pile of clay," the hope is that the complex molecular networks of cells can be constructed from these simpler "bricks."

Life as Information

Capturing the details of the great cobwebs of interconnected chemistry that spin within our cells is one way to model their behaviour but, if you see these biochemical reaction networks mapped out in detail, they are somewhat bewildering. As the Nobel laureate Paul Nurse complained, it is hard to see the wood for the trees. Nurse is interested in portraying life and cells in a different and more fundamental way, in terms of logic and information.[36] By this, he means much more than the current—and admittedly important—view of how information encoded in DNA is turned into proteins, but how information is used overall, both externally and internally, by a biological system, such as a human cell. "Information is central to biology. It was very common to talk about this in the 1950s and the 1960s, and then it got eclipsed by molecular biology."*

He believes that trying to understand cells at the chemical level is overwhelming and that it would be more fruitful to regard the cell as a glorified black box and understand how it deals with inputs and outputs of information. Then one could, for example, classify the collective cellular chemical reactions within the box as "modules" that may be involved in the dynamics of intracellular communication, including feedback loops, switches, timers, oscillators and amplifiers. Two classic examples include the viewing of DNA as a digital information storage device and the lactose operon (lac operon), a stretch of DNA code required for the transport and metabolism of lactose in *E. coli* and other bacteria. The lac operon can be viewed as a Boolean logic gate, a nonlinear negative feedback loop that activates itself upon the condition "lactose AND NOT glucose."

The hard part is working out how the multifarious and multitudinous molecular interactions in complex cellular biochemistry

* Paul Nurse, interview with Peter Coveney and Roger Highfield, September 25, 2021.

map onto logic modules. The reason it is not straightforward is because, as Nurse points out, living machines are not intelligently designed but have evolved to be overly complex, containing endless redundancy. Another reason the logic of life is messy is that its component molecules can unite in many different combinations.[37] The hope is that, when these modules are "wired up" by processes such as reactions, diffusion and cytoskeletal transport, where the cell's "skeleton" moves cellular components around, one can work out how the cell is organised in three dimensions. This also means that different information can be stored in different places and a wide variety of connections between logic modules can wax and wane through the diffusion of biochemicals during the life of a cell.[38]

This kind of higher-level description would free us from having to understand all the chemical processes within cells, just as a circuit diagram of resistors, transistors and so on frees us from having to know what all the electrons in an electrical circuit are actually doing. "We need to focus more on how information is managed in living systems and how this brings about higher-level biological phenomena," he says.[39]

That means working out how information is gathered from the environment, from other cells and from the short- and long-term memories in the cell, then integrated, rejected, processed or stored for later. "The aim is to describe how information flows through the modules and brings about higher-level cellular phenomena, investigations that may well require the development of new methods and languages to describe the processes involved." Like us, he sees the greatest hurdle is developing the necessary theory to lay bare the emergent processes of life.

Virtual Cell Experiments

Virtual cell models will ultimately give us the freedom to explore "what if" scenarios, observing the effects of altering the logic of life, or interfering with a protein using a drug molecule. Just as digital twins are now widely used in engineering, for instance, to see the effect of changing a structural component in a building, so genes can be mutated or removed from models to understand their effects.

At the University of Manchester, England, as one example, Henggui Zhang and colleagues have used heart cell models to understand the impact of gene mutations that alter a protein called SCN5A, involved in the generation of electrical activity. The models revealed the effects on the heart's pacemaker, the sinoatrial node, and the cause of a heart rhythm disorder called sick sinus syndrome, which can cause sudden death.[40]

A large number of molecules or signals inside cells affect how cancer cells lose their ability to attach to other cells and begin to migrate, as do the cancer's surroundings, from neighbouring cells to a web of proteins known as the extracellular matrix. There are too many factors for a single brain to make sense of, so a virtual cell was built at the Instituto Gulbenkian de Ciência and Instituto de Investigação e Inovação em Saúde in Portugal, providing more insights into how to hinder this deadly spread.[41]

While we are still working towards a high-fidelity description of virtual human cells, there has been important progress made at other levels of description, which—as Einstein suggested, and Noble, Nurse and others have echoed—are just as valid. These higher-level descriptions are needed to make sense of the body's defences, which range from earwax, tears and skin to swarms of cells—such as the T cells mentioned earlier—and endless factors, from antibodies to chemokines and histamine. This, of course, is the immune system, and over the past three decades there have also been attempts to create a virtual version, not least the T cell models we mentioned earlier.[42]

In 2022, this effort received a filip with the news that researchers had created the first map of the network of connections that make up the human immune system, revealing how immune cells communicate with each other. The study, led by the Wellcome Sanger Institute near Cambridge, England, details how the researchers isolated a near-complete set of the surface proteins that physically link immune cells and then used computational and mathematical analysis to create a map to show the cell types, messengers, and relative speed of each 'conversation' between immune cells. With this map, it is possible to follow the impact of different diseases on the entire immune system to, for example, help develop new therapies that can work with the body's defence mechanisms.[43]

Agent-Based Models

To model how cells interact, another kind of simulation is often used, a so-called agent-based model. You can think of each agent as an independent program that can decide its own fate, akin to individual cells in the immune system. Agent-based models can, for example, simulate the interactions between cells in basic states—once validated by experiments—to help settle matters of debate, or simulate emergent properties.

One example of their use is to explore the degree to which T cells are influenced by each other compared with encountering infected cells. Using this approach, one study by a team at the Institut Pasteur, Paris, and the University of Leeds found that T cells first have to be activated by a threshold number—85 cells—of another type of immune cell, so-called dendritic cells.[44] They speculate that this threshold is a safeguard mechanism against unwanted immune responses.

Another agent-based model was used to study sepsis, when the immune system spirals out of control.[45] The model suggested that true "precision medicine" would require a treatment to adapt the immune system based on each patient's individual response. The researchers concluded that computational modelling was necessary to aid the development of effective sepsis drugs.

Many features of the body can only be understood by studying how cells move, grow, divide, interact, and die. To simulate both the microenvironment (the "stage") and all the interacting cells (the "actors"), Ahmadreza Ghaffarizadeh, Paul Macklin and colleagues at Indiana University created PhysiCell, an open-source, agent-based simulator of cell behaviour in three dimensions.[46] Simulations of 500,000 to 1,000,000 cells are feasible on desktop workstations, many more using a high-performance computer.[47]

PhysiCell can be used, for example, to explore the complex cellular interactions that fuel the growth and spread of tumours, or to hone cancer immunotherapy, a promising but somewhat hit-or-miss treatment that realigns the immune system to cull cancer cells. Using two approaches—genetic algorithms, which harness evolution to explore possible solutions to a problem, and active learning, a kind of machine learning—the Cray XE6 Beagle at the Argonne National

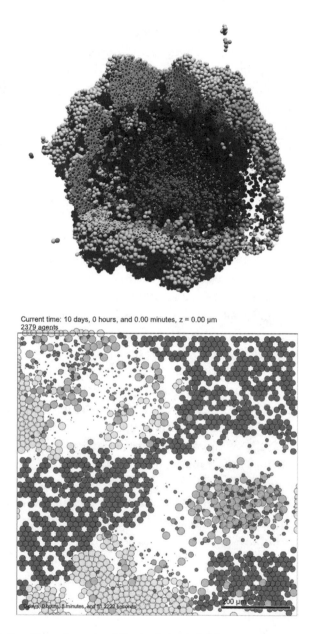

Current time: 10 days, 0 hours, and 0.00 minutes, z = 0.00 µm
2379 agents

200 µm

FIGURE 31. *Top*: Three-dimensional PhysiCell simulation (cutaway view) of a tumour being attacked by immune cells at 21 days. From "PhysiCell: An Open-Source Physics-Based Cell Simulator for 3-D Multicellular Systems" by Ahmadreza Ghaffarizadeh et al. (*PLoS Comput Biol* 14[2]. © 2018 Ghaffarizadeh et al.) *Bottom*: PhysiCell model of micrometastases (light cells) growing in tissues (dark), as part of a digital twin project. (Preliminary result adapted with permission from Heber Lima)

Laboratory was used to guide the PhysiCell model to reveal the factors that best controlled or killed simulated cancer cells.[48*]

More recently, Macklin, Michael Getz and colleagues led a multidisciplinary coalition of virologists, immunologists and modellers to create a model of innate and adaptive immune responses to a SARS-CoV-2 infection, showing the course of infection, from tissue damage and inflammation to T cell expansion, antibody production and recovery.[49] This immune model was subsequently adapted to drive a new generation of tumour-immune interaction models for digital twin simulations of micrometastases, breakaway cancer cells with the potential to develop into dangerous secondary tumours.[50]

Beyond individual cells, we need to simulate virtual tissues and organs. When it comes to cardiac medicine, for example, that means models to connect the geography of the heart with chemical processes at millions of points in the organ that create a beat. First, Denis Noble's cardiac model was extended in 1991 to a sheet of cells, showing how in cardiac arrhythmias spiral waves can originate, relatives of the spirals in the Belousov-Zhabotinsky reaction.[51] Since the late 1990s, the cardiac cell models have been extended to the whole organ. The first human ventricular model emerged in 1998, both for a normal heart and for one suffering heart failure, when one's personal pump becomes too weak or stiff to pump properly.[52] As we are about to see, many teams have been developing virtual hearts in recent years.

To deal with the whole human heart, these teams have to use partial differential equations, which are vastly more complicated than the ordinary differential equations in the very first heart cell model. For it to beat, a virtual heart has to couple partial differential equations to the ordinary differential equations used to describe its myriad component cells in a multiscale description that may also stretch down to the molecular level. You need more than data, theory and high-performance computers to create the wondrous pump in your chest. You also need to stitch together disparate kinds of theory in a multiscale, multiphysics model. With this fourth step, you can unite substance, form and function—in the guise of digital organ twins—within a virtual body.

---

* Readers can try this model themselves in a web browser at https://nanohub.org/tools/pc4cancerimmune.

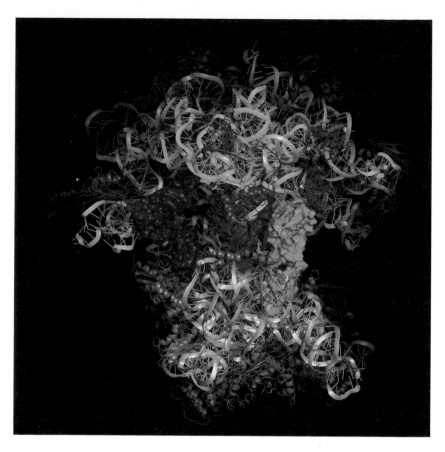

PLATE 1 (FIGURE 8). The ribosome. (Venki Ramakrishnan, MRC LMB)

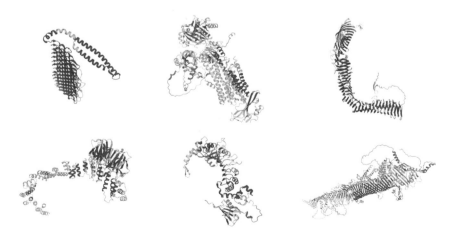

PLATE 2 (FIGURE 22). Proteins come in many shapes, which are critical for the way they work. (DeepMind)

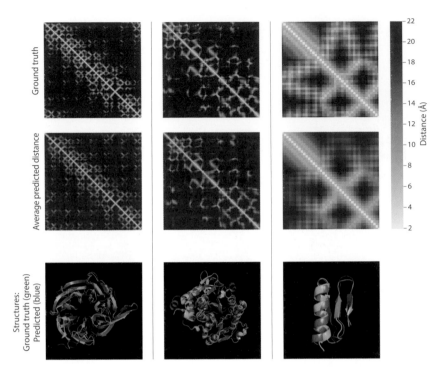

PLATE 3 (FIGURE 23). How to visualise the accuracy of AlphaFold's predictions: Here are the "distance matrices" for three proteins, where the brightness of each pixel represents the distance between the amino acids in the sequence comprising the protein—the brighter the pixel, the closer the pair. Shown in the top row are the real, experimentally determined distances and, in the bottom row, the average of AlphaFold's predicted distance distributions, which match well. (DeepMind)

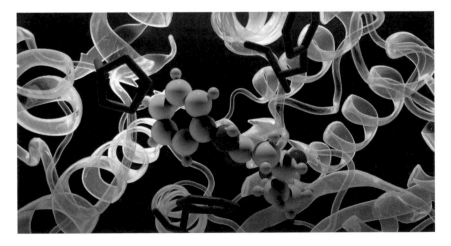

PLATE 4 (FIGURE 27). Simulation of a potential anticancer drug, binding to an enzyme linked with leukaemia. (CompBioMed and Barcelona Supercomputing Centre)

PLATE 5 (FIGURE 28). This painting depicts a coronavirus just entering the lungs, surrounded by mucus secreted by respiratory cells, secreted antibodies, and several small immune system proteins. The virus is enclosed by a membrane that includes the S (spike) protein, which mediates attachment and entry into cells. (Created by David S. Goodsell, RCSB Protein Data Bank)

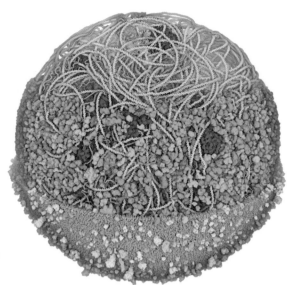

PLATE 6 (FIGURE 30).
3D model of a *Mycoplasma genitalium* cell. (Created by Martina Maritan, Ludovic Autin and David S. Goodsell, Scripps Research and RCSB Protein Data Bank)

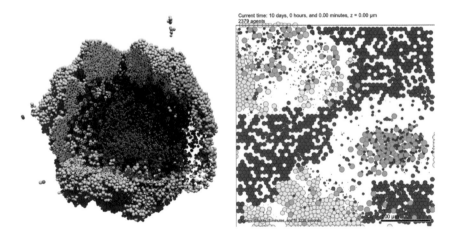

PLATE 7 (FIGURE 31). *Left*: Three-dimensional PhysiCell simulation (cutaway view) of a tumour being attacked by immune cells at 21 days. From "PhysiCell: An Open-Source Physics-Based Cell Simulator for 3-D Multicellular Systems" by Ahmadreza Ghaffarizadeh et al. (*PLoS Comput Biol* 14[2]. © 2018 Ghaffarizadeh et al.) *Right*: PhysiCell model of micrometastases (light cells) growing in tissues (dark), as part of a digital twin project. (Preliminary result adapted with permission from Heber Lima)

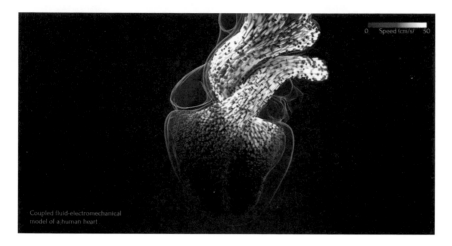

PLATE 8 (FIGURE 32). The Alya Red virtual heart. (CompBioMed and Barcelona Supercomputing Centre)

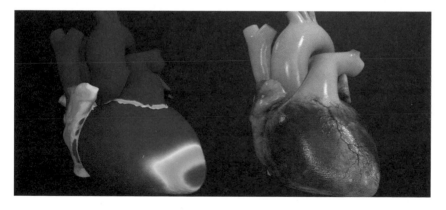

PLATE 9 (FIGURE 35). Living Heart Project. (Reproduced with permission by Dassault Systèmes)

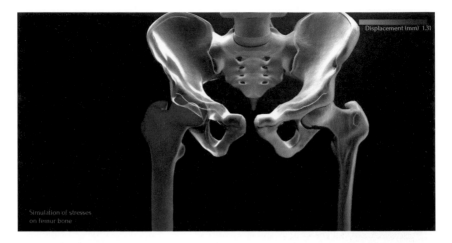

PLATE 10 (FIGURE 39). Simulation of stresses on a human femur. (CompBioMed and Barcelona Supercomputing Centre)

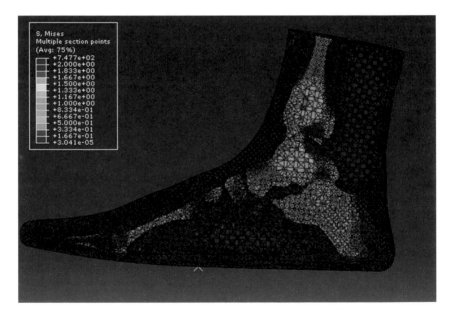

PLATE 11 (FIGURE 40). Digital orthopaedics. (Reproduced with permission by Dassault Systèmes)

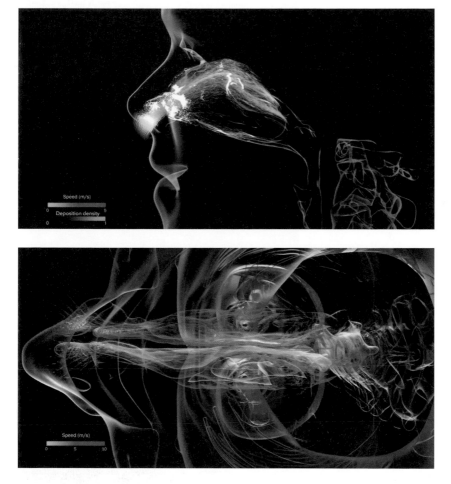

PLATE 12 (FIGURE 41). A virtual breath. (Credit: CompBioMed and Barcelona Supercomputing Centre)

PLATE 13 (FIGURE 48).
The Sycamore processor
mounted in a cryostat to
keep it cool. (© Google)

PLATE 14 (FIGURE 49). The quantum computer Jiuzhang 1.0 manipulates light using an arrangement of optical devices. (© The University of Science and Technology of China)

How to Create a Human Heart

There are no limits to what science can explore.

—Ernest Solvay

"A Witches' Sabbath" is how Albert Einstein once described the first Conseil Solvay, an august gathering in 1911 that was named after its main supporter, the wealthy Belgian chemist and industrialist, Ernest Solvay (1838–1922).[1] The inaugural Solvay meeting in Brussels more than a century ago marked the first international conference dedicated to physics, one that would change the future of the field because it put theoretical physics on the map.

Attended by Einstein along with Marie Curie, Henri Poincaré, Ernest Rutherford and other luminaries, the delegates wrestled with the revolutionary implications of quantum theory, born only a few years beforehand. At that time, the theory's insights into the workings of the atomic world were as perplexing as they were illuminating. As we will see in Chapter Nine, there are still arguments over how to interpret quantum mechanics. But when it comes to the Conseils Solvay, or Solvay Conferences, they have grown in stature over the past century to become among the best known in all of science.

Where better to discuss the creation of Virtual You? In April 2016, we both gathered with some 50 researchers from around the world at a Solvay meeting on the campus of the Université Libre de Bruxelles to discuss a range of topics that, to a casual observer, seemed to have little in common. There was our critique of big data and AI, which we outlined earlier,[2] along with papers that ranged from molecular and genomic studies to work on cell, organ and organism, and from modelling viral infectivity to creating virtual arteries and simulating blood flow within the brain.

However, underpinning our disparate discussions was a single problem, one central to attempts to simulate features of the world and at the heart of the fifth step required to create Virtual You. How is it possible to come up with a single model that can capture phenomena at broad and diverse scales in space and time, and at the interface between many different disciplines, such as physics, chemistry, materials science, molecular biology and medicine?[3]

The human body presents particular challenges, being a chemical machine of such boggling complexity that it is not straightforward to scale up models from the molecular level to that of the cell, let alone to organs or the whole human level, and vice versa. The challenge is to create a multiscale model and, when you draw on different pedigrees of theory—for instance, to handle electrical and mechanical processes—a multiphysics model too.

Wiring up these processes is crucial. The workings of the body abound with nonlinearities, where feedback damps some processes down (think of a thermostat in household central heating) and winds other things up (think of howlround, or feedback, when a microphone is too close to a loudspeaker). There are feedback loops going up and down in scale, from cell to organ to body, where different kinds of physics come to bear.

Understanding feedback is crucial because your life depends on it. One example can be found in the stable, relatively constant internal environment within your body. Homeostasis regulates body temperature and the concentration of various ions in your blood, along with glucose and pH. Maintenance of homeostasis involves negative feedback loops so that, if your temperature is too high, nerve cells with endings in your skin and brain will trigger sweating. To understand how the nonlinear whole of the body emerges from its physiological, cellular and molecular parts is central to the problem of how to create Virtual You.

Scaling the Multiscale

The key issue in multiscale and multiphysics modelling, whether used to simulate a nuclear explosion or the detailed workings of the human body, is to find a way to connect different theories that govern different regimes and to model methods that wring the most value out of available computer power.

Some multiscale issues revolve around how fine-grained a simulation needs to be. There is tension between large-scale models that are easy to run on a computer, but not particularly accurate, and detailed microscale models that are accurate, but grindingly inefficient. When it comes to weather forecasting, for example, running a model that has a resolution down to a mesh consisting of squares 10 km on each side will miss a lot of detailed weather and contain other inaccuracies, but will be much quicker than a model of 1 km resolution, even though the underlying mathematics remains the same.

Multiscale thinking can provide solutions. The best of both worlds is "adaptive mesh refinement," or AMR, where the coarse mesh is initially used and the computationally expensive fine one reserved for use in the parts of the simulation where there is complex geography or behaviour, as for instance, when you need a more detailed glimpse of the weather. In that way, maximum computer power is only used to zoom in on a feature of interest, such as a town that is prone to flooding.

This fifth step has also been crucial when it comes to modelling cancer, now a regular feature of the annual Supercomputing Conference, held in the US. In Chapter Five, we discussed modelling the dynamics of RAS proteins, a family of proteins whose mutations are linked to around one-third of all human cancers. The team from Lawrence Livermore National Laboratory, along with scientists from Los Alamos National Laboratory, the National Cancer Institute and other institutions, began with a simulation of the impact of a lipid membrane on RAS proteins at long timescales and incorporated a machine learning algorithm to select which lipid "patches" were interesting enough to model in more detail. The result is MuMMI, massively parallel multiscale machine-learned modelling infrastructure, which scales up efficiently on supercomputers such as Sierra and Summit.[4]

Multiscale modelling could help find the right treatment at the right time and the right dose for the right patient. Because the use of genetic markers to help customise cancer therapies has proved relatively disappointing in real-world applications, Peter is involved in another multiscale cancer initiative to blend machine learning (to find correlations among big data) and multiscale mechanistic modelling (to find causal relations between data). As discussed in Chapter Four, this Big AI approach is promising because machine learning in isolation ignores the fundamental laws of physics and can result in ill-posed problems or nonphysical solutions, while mechanistic modelling alone often fails to efficiently blend big data from different sources and varying levels of resolution.[5,6]

Other multiscale issues revolve around connecting the physics of the molecular world with that of the world we appreciate with our senses. Earlier, we discussed the nineteenth-century origins of this endeavour, in Boltzmann's work on statistical mechanics. This effort to link the micro with the macro continues. A Nobel Prize in 2013 acknowledged one key advance in multiscale modelling in chemistry made by Martin Karplus of the Université de Strasbourg in France and Harvard University, Michael Levitt of Stanford University and Arieh Warshel of the University of Southern California. Decades earlier, they had combined the classical physics that rules the everyday world with quantum physics of the microscopic realm that emerged just before the first Solvay meeting. While the classical theories of Newton predict the movement of limbs and cannonballs, you need the quantum theories of Erwin Schrödinger, Paul Dirac and others to make predictions of events at the atomic level.

They found a practical way to blend the best of both worlds. In 1967, Levitt and Warshel wrote a computer program that used traditional Newtonian physics to describe the resting structure of large molecules, based on their amino acid sequences, then extended the algorithm to include quantum physics. In 1972, working together at Harvard, Karplus and Warshel constructed a program that could calculate the spectra of complex molecules—the unique pattern of light they give off when excited—by blending classical and quantum chemical approaches.[7] In 1976, they published the first model of an enzymatic reaction and, in this way, laid the foundations for some of the powerful computer programs that are used today to understand and predict chemical processes and for drug design.[8]

There are other ways to deal with feedback between the classical and quantum domains. One example can be found in modelling a flowing fluid, where a part of the simulation is treated with classical molecular dynamics, so that the behaviour of component molecules is simulated while the rest is solved using Navier-Stokes equations that are oblivious to the molecular turmoil in a fluid. Here each level of description must be consistent, so you need to ensure mass, momentum and sometimes energy are conserved. Peter was among the first to show how this could be done in the early 2000s.[9]

Modelling such "nonhierarchical coupled fluid models" is not easy: if there is a flux of mass from the region described by continuum physics into the domain of atoms and molecules described by molecular dynamics, then virtual molecules have to be introduced into the simulation on the fly to capture these ebbs and flows.

Multiscale modelling has to deal with practical problems too. One is "load imbalance," where the simulations invest so much time at the microscopic/fine-grained level that a bottleneck results, and the rest of the large-scale simulation is left "kicking its heels" and cannot move on to the next time step. One popular way to balance load is to replace the fine-grained, compute-intensive part of the model with a surrogate, a neural network or other AI that has been trained to behave in the same way. As ever, for reasons outlined earlier, these surrogate models are only as good as the data used to train them. Of all the multiscale challenges within the human body, one of the most important—given the global burden of cardiovascular disease—is modelling the human heart.

Multiscale Hearts

It should come as no surprise that the heart is one of the stars of the effort to create a virtual human, given the enduring symbolism of this magnificent organ, from its link to the soul and intellect in ancient Egypt to notions of *erotas* (romantic love) and *agape* (spiritual love) in ancient Greece, to emojis of health, joy and pain that flit between mobile devices today.

This wondrous pump beats 100,000 times every day to supply oxygen and nourishment to your body, regulate your temperature, carry the cells and antibodies that protect you from infection and remove carbon dioxide and waste products, all via a 60,000-mile-long network of vessels that carries around 4.5 litres of blood. Together, they comprise the cardiovascular system.

Although it weighs less than a pound, your heart shifts more than 6000 litres of blood daily and is powerful enough to create a gory spurt if you are unlucky enough to sever an artery. It has a degree of autonomy too. Prehistoric hunters would have found that a heart does not need the body to tell it when to pump—if you remove a heart from an animal, it will continue to beat for a while. A century ago, the French surgeon, Nobelist and biologist Alexis Carrel (1873–1944) showed that a chick's heart tissue, when nourished with blood plasma, is able to pulsate for months.[10]

The heart's pumping action depends on the coordinated contraction of its four chambers and that in turn depends on rhythmic electrical impulses. Each squeeze is regulated by specialised muscle cells found in a clump of tissue in the organ's right upper chamber (atrium), variously known as the sinus node, also the sinoatrial node or SA node, which sends out a pacemaking signal, the subject of the very first cardiac models. The signal travels along conduction pathways to stimulate the two upper chambers of the heart (atria) first, as the impulse travels from the sinus node to another node, the atrioventricular node (also called the AV node). There impulses are briefly slowed, then ripple down the "bundle of His" (pronounced *hiss*)—a bunch of cardiac muscle fibres named after a Swiss-born cardiologist—which divides into right and left pathways, called bundle branches, to stimulate the lower chambers of the heart (ventricles) via thin filaments known as Purkinje fibers, which are named after the Czech anatomist who, among other things, helped establish the uniqueness of fingerprints.

The heart's walls dilate as blood flows in (diastole) and contract again as the blood flows out (systole). The overall movement is caused by millions of tiny contracting units within muscle fibres, the sarcomeres, molecular machines that form the smallest functional unit of skeletal and cardiac muscles. They are packed with tightly interacting proteins, notably titin, the largest protein in the human body, along with actin and myosin.

When a muscle is triggered to move, myosin "motors" move along actin tracks to bring neighbouring sarcomeres closer together.[11] This sliding of actin and myosin filaments creates enough force to pump blood. Meanwhile, the titin protein scaffold works like a spring, developing a restoring force during the stretching of the sarcomeres, a little like a rubber band.

If we are to create a digital simulacrum of the heart, its twin has to capture how its beat emerges from many twitching heart muscle cells, the protein machines that help contract each cell, the electrical waves that trigger a squeeze of cardiac tissue and the orchestrated surges of ions (electrically charged atoms) in and out of cells, along with the resulting surges of blood through the heart's four chambers.

Brief History of Virtual Hearts

In human embryos, the heart is the first functional organ to develop and starts beating spontaneously only four weeks after conception.[12] The effort to simulate the human heart in a computer has taken much longer, dating back more than half a century, to the inspiration of the Hodgkin-Huxley model.

To create the algorithms on which to simulate a heart, there are various ways to incorporate patient data. Scans of a heart can be digitized in the form of a polygon mesh, when collections of vertices, edges and faces are used to define the shape of an object. In this way, the structure and shape of the heart can be drawn as meshes, providing a scaffold for the heart cell models dealt with in Chapters Three and Five.

However, rather than model individual cells, which would demand far too much computer power, modellers begin with blocks of cells, tantamount to little scraps of cardiac tissue, a familiar approach used by engineers, who design cars, bridges, aircraft and more using the "finite element method": a continuous material, whether flesh or a girder, is regarded as a collection of small regions (finite elements) linked together through points (nodes) on their boundaries. In that way, the geometry of a complex object like a heart (or indeed a car) can be divided into smaller problems that can be more easily conquered.

This reduces a great swathe of hundreds of thousands of cells in cardiac tissue to two partial differential equations, the equations that mathematicians use to capture change in space and time at larger, "engineering" scales. This "continuous model," known as the bidomain model, has to be converted into another form of discrete model once again—dividing the heart up with a mesh—to conduct simulations. As a result, the two partial differential equations (which are hard to solve) are splintered by the mesh into tens of thousands of tractable algebraic equations that hold increasingly true, the smaller the increments of space and time.

Working with Rai Winslow in the US, Denis Noble started work on a multicellular tissue and whole organ model of the heart in 1989, using a supercomputer at the US Army High Performance Computing Research Center at the University of Minnesota. Called the Connection Machine CM-2, this influential parallel processor was the brainchild of the American computer scientist, "imagineer" (a position he later held at Disney Inc.) and inventor, Danny Hillis, and made by Thinking Machines Inc., based in Waltham, Massachusetts. The CM-2 harnessed 64,000 very simple processors and, as Noble explained: "You could therefore map a block of tissue onto the processors themselves, in the simplest case by assigning a cell model to each processor, so generating a block of tissue corresponding to 64,000 cells."

As virtual heart tissue was being created on the CM-2, Noble visited Auckland in 1990, where he encountered Denis Loiselle, one of the founding members of the Auckland Bioengineering Institute. "I am an electricity man, he is a mechanics man, and it didn't take long for he and I to see what was missing from the primitive cardiac electrophysiological models that I had arrived with," recalled Noble. "He asked an apparently simple question. 'Where is the energy balance in your models?' Well, there wasn't any! This was the beginning of connecting metabolism to electrophysiology." The incisive and patient questioning from Loiselle would later spur Noble on in Oxford where, with Kieran Clarke and Richard Vaughan-Jones, he would model ischaemia, when blood flow becomes reduced or restricted in the heart.

In the early 1990s, Noble was blending more biochemical processes and nutritional processes into his cardiac model, notably ATP, or adenosine triphosphate, the energy currency of life that drives the contraction of your muscles and the beating of your heart, along with the functioning of your brain. Rather like energy can be stored in a bent spring, so ATP locks it up within its energy-rich phosphate bonds. Snap one of those chemical bonds and energy is released to do useful work in a living cell.

Around this time, the idea of what today is called the Physiome Project also winked into life.[13] Noble, along with Peter Hunter in Auckland, conceived of a framework for modelling the human body, using computational methods that incorporate biochemical, biophysical and anatomical information on cells, tissues and organs. "The vision that I encountered during those few weeks as the Butland Visiting Professor was simply mind-blowing," recalled Noble. "I see Peter Hunter as the intuitive visionary of the whole project."

By the time Noble was working with Winslow, Hunter was digitizing dog hearts by progressively shaving off thin layers and imaging the fibre orientations over the entire exposed surface, then taking two-dimensional images that could be built up into a 3D digital version.[14] Noble recalled that when he witnessed the process during his visit, he was struck by how painstaking it all was, and how useful too: "The idea was that this database could eventually be combined with equations for the mechanical, electrical and biochemical properties of the cells in each region of the heart to produce the world's first virtual organ."[15]

The pair had first collaborated when Hunter was in Oxford in the 1970s, studying for his thesis on heart mechanics. Noble recalled: "One day he was looking over my shoulder while I was discussing a particular mathematical problem on excitable cells and just calmly said, 'I think the solution must be X.' He was right."

The problem in question was whether the Hodgkin-Huxley equations could be solved analytically, with a definitive equation that would not require laborious numerical computation of differential equations for many different cases.* The advantage of such an analytical solution, as mentioned earlier, is that it has great generality, whereas numerical integration of differential equations gives a solution only for each particular case. For example, the equation can

---

* Denis Noble, email to Roger Highfield, May 28, 2021.

reveal the way in which the speed of conduction of a nerve (or indeed a heart) impulse depends on the radius of the fibre, its internal resistance, membrane capacitance and the gating properties of the sodium ion channel.[16] That insight was eventually proved rigorously and became established.

Despite this progress with analytical solutions, to create the first heart models that dealt with cardiac tissue structure and mechanical function would take Hunter another decade of research in New Zealand, working with Bruce Smaill and students Andrew McCulloch, Paul Nielsen and Ian LeGrice.[17] These models were supplemented with virtual coronary blood flow, by Hunter and his student Nic Smith in Auckland,[18] and then with electrical activation processes in another collaboration with Denis Noble at Oxford.[19]

As he continued to work with Rai Winslow on using a Connection Machine, Noble was now able to reproduce the behaviour of a block of cardiac tissue and, in this way, could solve a puzzle concerning the origins of the heartbeat. The sinus node is the pacemaker, and Noble was keen to understand the origin of the impulse. At that time, the thinking was that the impulse usually starts at or near the centre of the node but, to their dismay, their computations showed precisely the opposite, suggesting the impulse started at the periphery.

Unsettled by this finding, Noble nonetheless presented the results at a meeting as an example of what could be achieved in multilevel simulation using the Connection Machine, adding that, of course, there was a problem. He need not have apologised. "The result is correct," replied one member of the audience, Mark Boyett of the University of Leeds, who had already shown that if you carefully cut around the sinus node to isolate it from the atrial tissue, the impulse does indeed start at the periphery.[20]

Rapid progress followed in animating whole hearts. Two decades ago, an entire dog or rabbit heart could be mimicked with several million grid points, so that a virtual beating version could be nourished with virtual sugar and oxygen. Even though a simple virtual organ, it was already capable of providing insights into clinical trials.

These models were always designed for use with patient data in mind, but this feature was developed much further in Auckland by Andrew McCulloch and Nic Smith, and more recently by Steven

Niederer, who went on to work in Oxford and now leads a group at King's College London that focuses on customising heart models using MRI and other clinical imaging data. We will return to these bespoke virtual hearts.

The first cellular model for electrophysiology that Denis Noble based on ordinary differential equations has spawned four parallel lanes of digital heart twin activity. The primary lane remains the data harvested on the heart by cardiologists, giving deeper and richer insights into how it works, and when it goes awry in arrhythmias. The second has continued to model this evidence at the cellular level, elaborating Noble's single cardiac cell model, as we have seen in the work of Yoram Rudy at Washington University in St. Louis and Blanca Rodriguez at Oxford.

But, of course, this will miss the emergent properties of many cells, for instance, how many cooperate to create a beat. The heart muscle, or myocardium, consists of something like 10 to the power of 10 cells, that is 10,000,000,000 cells. To model in a bottom-up fashion the electrical propagation through this colossal number is a tall order. Instead, the third lane of activity divides the whole organ into a mesh and animates it with partial differential equations. The most recent, fourth, lane of activity is to simulate longer-term changes in the heart—remodelling—which take place after, for example, a pacemaker implant.

There is, of course, traffic between these lanes of activity. The second and third cross when the mesh, which models the heart down to a few tenths of a millimetre, is informed at each node by insights from simulations run with a more sophisticated cardiac cell model that resolves processes down to the ion channels that allow charged atoms to flow in and out of heart cells. Since Noble worked on the first heart cell model on a primitive computer in Bloomsbury, the sophistication of virtual hearts has soared.

Heart Twins

In his novel *Origins*, the fifth installment of Dan Brown's Robert Langdon series, the hero finds himself in Barcelona while searching for

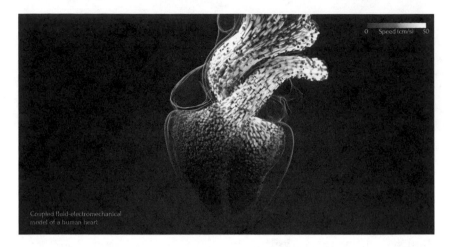

Coupled fluid-electromechanical
model of a human heart

FIGURE 32. The Alya Red virtual heart. (CompBioMed and Barcelona Supercomputing Centre)

a cryptic password and evading the clutches of an enemy. The Barcelona Supercomputing Center features in the novel several times, along with MareNostrum and Alya Red, a multiscale model of the heart that has grown from blending macroscopic, continuum models of blood flow and mechanics with models of the electrical activity within cells. This inspirational multiscale mix of electrophysiology, fluid mechanics and mechanics beats just like the real thing.

This is one of many models to emerge over the past decade that are providing better and better overall approximations to the whole heart, informed by advances in heart imaging with MRI and CT scans.[21] At Washington University in St Louis, Yoram Rudy is using whole organ simulations to predict electrocardiogram patterns in lethal arrhythmias. There is also the Living Heart Project, which creates digital twins of patient hearts, developed by Dassault Systèmes Simulia Corporation and others,[22] along with studies by Gernot Plank at the Medical University of Graz, Austria, and research by Toshiaki Hisada at the University of Tokyo.[23]

Mariano Vázquez, who worked with us to create the spectacular *Virtual Humans* film we premiered in the Science Museum, is among the 50 researchers in the Alya development team at the Barcelona Supercomputing Center. Since 2004, when they did their doctorates on modelling fluid flows with the same supervisor, he and Guillaume

Houzeaux have developed software for simulating a host of multiscale, multiphysics problems, from cloud formation to the way water flows around whales. They named their software after stars, Alya in this case, and began to customise it, so Alya Red was used for biomedicine, Alya Green for the environment and so on. A spinoff company, Elem Biotech, developed the model for biomedical applications, for instance, to show that the fine particles of drugs from an inhaler end up in the right place in the body, and worked with researchers at Washington University to model the contractions during childbirth, which, like the heart, are electromechanical.

At the time of writing, their Alya Red heart model consisted of around 100 million patches of cells, each of which is described by around 50 equations in terms of 15 variables (sometimes more, depending on the model's complexity), such as electrical activity, displacement, pressure and velocity.* Just as in weather forecasting, subdividing (discretising) the heart into a series of smaller, more manageable pieces produces a set of billions of coupled finite difference equations that must be solved to run the model. Compared with the thousands of millions of numbers that are required when running the Alya Red model, relatively few parameters are needed to customise the digital heart for a patient. Details of the heart of interest—male, female, young, old, obese, diabetic and so on—come from measurements of human hearts. Some are collected by the University of Minnesota's Visible Heart Lab, for example.

Typically, it can take 10 hours to simulate 10 heartbeats. Leaving aside how, in an ideal world, we would like to simulate the human heart much faster than the real thing, these simulations are still revealing. A cutaway can show, for example, how a failing heart loses its ability to pump, or that a dangerous arrhythmia is caused by heart drugs. The flows in a normal heart can be made easy to visualise in the simulation by depicting them as bundles of vivid colour, with red, orange and yellow used to signify the strongest flows. By comparison, the tracks of flow within diseased ventricles, no longer properly synchronized, are revealed as sluggish blues and greens.

---

* Mariano Vázquez, interview with Roger Highfield, September 4, 2021.

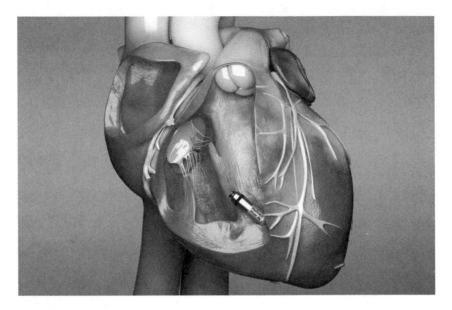

FIGURE 33. The Micra pacemaker. (Reproduced with permission by Medtronic)

Even with a basic virtual human heart, much can be done to aid the design, development and manufacture of medical devices, which is the subject of a collaboration under way between the company Medtronic and the Barcelona Supercomputing Center. Their simulations show how a traditional pacemaker, with its electrodes threaded through veins, can restore the pumping action of a failing heart. They can help position the pacemaker and fine-tune its electrical stimulus. They have also helped Medtronic to model the effects of an innovative pacemaker, the Micra, a tiny cylinder that is inserted into the ventricle itself, showing the best place to put it to ensure it works, to investigate how to attach it without harming the heart, and to study the resulting flows of blood.

The next step is to create a digital twin of a patient's own heart. Working with Gaby Captur of the UCL Institute of Cardiovascular Science, supported by the British Heart Foundation, customisation is being carried out with the help of MRI and echocardiography data from 500 Britons born in 1946. More heart data is gathered at lower resolution by other teams, such as the UK Biobank, a remarkable global source of data on 500,000 people. The Barcelona team often creates a synthetic heart population to work out the most important parameters, then plays with around 100 variables to fashion the best-matching virtual heart to a patient, whatever their age, gender and so on.

The team began to create these bespoke models to understand "left bundle branch block," a common heart conduction abnormality that delays activation of the left ventricle, so that the left ventricle contracts later than the right.* "Those applications for individual patients are coming," Vázquez says. In a strange way, they are becoming a victim of their own success, however. Some people have complained to him that the simulations are too beautiful to be real, indeed that they look like "science fiction."

The heart also still commands the attention of most people in Peter Hunter's team in Auckland. By simulating the electrical activity in virtual hearts that have been customised with imaging data of the real thing, then comparing these predictions to high-resolution maps of cardiac electrical activity, new insights are emerging into heart rhythm abnormalities. Jichao Zhao is studying atrial fibrillation, the most common heart rhythm disturbance. *Fibrillation* derives from the Latin for a "bag of worms," a vivid reference to how the muscle wall of the heart's upper chambers (the atria) squirms in an uncoordinated way, so the atria no longer pump properly. Working with Vadim Federov and his team at Ohio State University, Zhao based his modelling on MRI measurements of the heart of a 63-year-old woman who had died in a car accident. In this way, they could lay bare how structural remodelling of the atria in failing human hearts makes them prone to fibrillation, as scarring (fibrosis) crinkles the thin atrial wall.[24]

This research also helped them to work out the ideal place in the heart to carry out ablation, where heat or extreme cold is used to scar or destroy tissue that fosters incorrect electrical signals, causing an abnormal heart rhythm. They are using a similar approach to understand how changes in the ventricles contribute to ventricular arrhythmia and sudden cardiac death. We are approaching the day when patients will routinely benefit from drawing on insights from a sophisticated personalised virtual version of their own heart.

* Mariano Vázquez, email to Roger Highfield 23/01/21.

Precision Cardiology

The eventual aim of the work in Auckland, Barcelona and elsewhere would be not only to make twin hearts accurate at every scale, from each beat to the idiosyncratic architecture of heart chambers to details of cellular metabolism, shape of ion channels and so on, but also to realise all the patterns and rhythms of the actual heart of an individual patient. In cardiovascular medicine, an array of virtual human methods have paved the way for the dawn of what Mark Palmer of Medtronic,* among others, calls precision cardiology. These models, based on computationally less intensive ordinary differential equations, are allowing heart twins to be customised.

At King's College London, for example, the Personalised *in Silico* Cardiology Consortium led by Pablo Lamata has shown how linking computer and statistical models can improve clinical decisions: "The Digital Twin will shift treatment selection from being based on the state of the patient today to optimise the state of the patient tomorrow."[25] Personalised virtual heart models have been created with the help of cardiac MRI by a team led by Reza Razavi at King's to predict tachycardia,[26] while a group at the University of Heidelberg, Germany, turned a range of patient data into a multiscale, multiphysics model of cardiac function to manage heart failure, when the heart cannot keep up with its workload.[27] In Graz, Austria, Gernot Plank's team is creating digital heart twins to improve the success of implanting pacemakers.† In a study of a dozen patients, his team found that they could create an anatomical heart twin in less than four hours from MRI images.[28] Most important of all, they could reproduce the electrocardiogram that marks out the heart's rhythm and electrical activity. Meanwhile, in Russia, Olga Solovyova of the Institute of Immunology and Physiology in Ekaterinburg is developing personalised models by comparing virtual ventricles with the real thing by comparing simulated and real electrocardiograms.[29]

---

* Mark Palmer, "Digital Twins in Healthcare," CompBioMed Conference, September 17, 2021.

† Gernot Plank, "Automating Workflows for Creating Digital Twins of Cardiac Electrophysiology from Non-invasive Data," CompBioMed Conference, September 16, 2021.

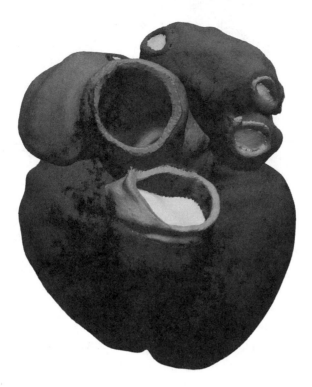

FIGURE 34. Digital twin models will predict future heart health. (Reproduced with kind permission by Cristobal Rodero, Pablo Lamata, and Steven Niederer)

Simulations can ensure that cardiac doctors use the right kinds of implants. Transcatheter aortic valve implantation is used to treat severe aortic stenosis, a heart valve disease that is one of the major causes of cardiovascular death worldwide. The disease narrows the heart's aortic valve, preventing it from opening fully, so that blood flow is blocked or reduced and the heart is forced to work harder. Stenosis generates all kinds of symptoms, from shortness of breath and chest pain or tightness, to fainting, dizziness, and heart pounding. Computer simulation by a team at the Erasmus Medical Center, Rotterdam, has helped doctors find the valve size and depth of implantation to suit the patient.[30] The same goes for operations to implant stents, tubes of scaffolding that are inserted into arteries to hold them open, where manufacturers of medical devices can put a new design through its paces in a virtual artery.[31]

Modelling can help do away with invasive tests. HeartFlow Analysis, a cloud-based service, enables doctors to create personal heart models from computerised tomography scans of a patient's heart, when computers stitch together a series of X-ray scans into detailed images. These CT scans can be used to construct a fluid dynamics model of the heart to reveal how the blood pumps through coronary vessels, helping to identify coronary artery disease and how to treat a blockage. Without this analysis, known as HeartFlow FFRct (fractional flow reserve computed tomography), doctors would need to perform an invasive angiogram.[32]

In France, Dassault Systèmes has conducted *in silico* clinical trial experiments within its Living Heart Project to create a cohort of "virtual patients" to help test a synthetic artificial heart valve for regulators, working with the US Food and Drug Administration.[33] Modelling and simulation can also inform clinical trial designs, support evidence of effectiveness, identify the most relevant patients to study and assess product safety, such as pacemaker leads and stents. Doctors can get even deeper insights by examining virtual reality and three-dimensional models of a patient's heart.

Compellingly, in some cases, clinical trials based on modelling and simulation have produced similar results to human clinical trials.[34] This represents an important milestone for Virtual You because, until recently, regulatory agencies have relied on in vitro or in vivo experimental evidence to authorise new products. Now agencies have started to accept evidence obtained *in silico*, thanks to the American Society of Mechanical Engineers "Verification and Validation in Computational Modeling of Medical Devices" technical committee, the Medical Device Innovation Consortium, also in the US, and the European Commission–funded "Avicenna: A Strategy for *in Silico* Clinical Trials."

At the time of writing, many efforts are under way to study *in silico* medicine: the VPH Institute in Belgium; National Simulation Resource Physiome at the University of Washington in the US; Insigneo Institute for *in Silico* Medicine at the University of Sheffield in the UK; Interagency Modeling and Analysis Group (IMAG), which coordinates US government research; and the Global Center for Medical Engineering and Informatics at Osaka University in Japan.

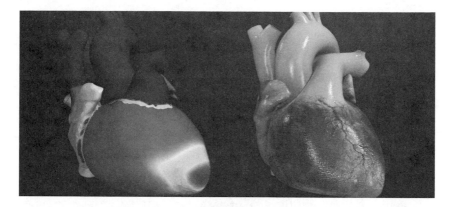

FIGURE 35. Living Heart Project. (Reproduced with permission by Dassault Systèmes)

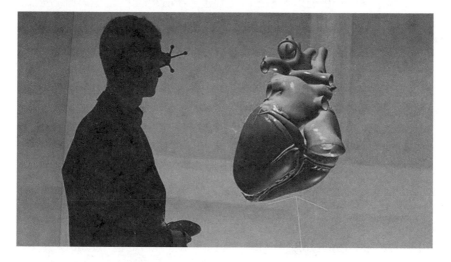

FIGURE 36. Using immersive methods to study the Living Heart model. (Reproduced with permission by Dassault Systèmes)

Simulations have helped to get to the bottom of enduring problems in heart medicine. One project, by an American–South African team, used a multiscale model of heart failure to link the decline in pumping action to the number of sarcomeres and the length of myocytes, muscle cells.[35] Others have studied why ablation is not always a successful way to treat persistent atrial fibrillation, the commonest heart rhythm disturbance, one that boosts the risk of stroke. In the treatment, doctors use ablation to isolate the pulmonary vein in a part of the left upper chamber. With the help of a computer model, Edward Vigmond and colleagues of the Electrophysiology and Heart Modelling Institute at Bordeaux Université came up with a way to work out the likelihood of success for a given patient.[36]

Meanwhile, at Johns Hopkins University in the US, a team led by Natalia Trayanova is also creating personalised digital replicas of the upper chambers of the heart, based on image data, to guide treatment of patients suffering from persistent irregular heartbeats, by the carefully targeted destruction of tissue through ablation.[37] A blend of cardiac MRI imaging and modelling has already been used by Trayanova and colleagues to work out the risk of arrhythmia in patients who have suffered heart attacks.[38] These models can help judge if a patient should have a defibrillator implanted to deliver beats and shocks if a dangerous heart rhythm develops.

To create a high-fidelity digital heart requires more than anatomical personalisation, however. To make sense of the data, you need to understand the model's uncertainty and sensitivity to define the "physiological envelope" of a person, verify and validate these models to build trust in their predictions and constantly update these models with data as a person goes about their life.[39]

One practical factor limiting the creation of digital heart twins is computer power. Real-time modelling of cardiac dynamics demands supercomputers to solve the billions of differential equations that capture the complex electrophysiology of interconnected cells, each of which is described by up to 100 differential equations. However, Flavio Fenton and colleagues from the Georgia Institute of Technology have harnessed graphics processing units—GPUs—and software that runs on standard web browsers to shift high-performance cardiac dynamics simulations onto less costly computers, even high-end smartphones.[40] They point out how a modern consumer-level GPU through parallelization can solve up to 40,000,000,000 ordinary differential equations per second. This advance could enable doctors to use 3D modelling data to design specific therapies or prevention strategies for their patients, and help scientists study a particular drug's effect on heart arrhythmias.

Steven Niederer is amazed at the progress that has been made since his first stumbling simulations of rat hearts while doing his doctorate in Oxford. Today, working at King's College London, Niederer is using image-based cardiac models to guide where to put pacemakers on hearts, and creating virtual cohorts of patient hearts and atria *in silico* to test therapies and devices.* A glimpse of the power of this approach can already be seen in one project to track changes in the spines of patients with scoliosis, a common cause of deformity, and use neural networks to predict the impact of surgery.[41] He is particularly struck by how commercial cardiac simulation software is now available from the likes of Dassault and the American company ANSYS, signalling the transition of this technology from the efforts of a few specialists to more routine use in university and commercial laboratories.

However, though computer models have enormous potential in cardiology,[42] that will only be realised when digital twin organs are routinely updated with a patient's data. The current generation of heart twins are, in a sense, disposable, says Niederer: created for a patient to help with a particular cardiac problem at one particular time. Although these contribute to the well of scientific knowledge, and are now of interest to doctors and manufacturers of medical devices, the longer-term hope is that these heart twins will guide the care of patients over their entire lifetimes.

We must be wary of hyperbole, however, often where dazzling cardiac graphics belie real understanding. As Gernot Plank put it: "There are simply too many in our community making bold promises, with an impressive expertise of turning significant funding opportunities not into any tangible outputs but into even more bold visions." Plank cites the Linux creator Linus Torvalds, the Finnish-American software engineer, when he was asked about his vision for the field: "I believe more in having passion. I think really caring about what you do is way more important than having this vision about the golden future that you want to reach."† Though digital twin hearts are already beating in clinics, they are likely to take longer to enter routine use in surgeries and hospitals than their most passionate advocates predict.

---

* Steve Niederer, interview with Roger Highfield, August 10, 2021.

† Gernot Plank, email to Peter Coveney, October 4, 2021.

Beyond the Digital Heart

We can now begin to think about how to connect a virtual heart with digital twins of the arteries and veins and finer plumbing that make up the cardiovascular system. In the 1950s, the American physiologist Arthur Guyton (1919–2003) studied the heart and its relationship with the peripheral circulation, challenging the conventional wisdom that it was the heart alone that controlled cardiac output. The resulting Guyton model could be regarded as the first "whole-body," integrated mathematical model of a physiological system, revealing the relationship between blood pressure and sodium balance, and the central role of the kidneys in controlling blood pressure. One recent study by Randall Thomas at the Université Paris-Sud, Orsay, put the model through its paces on a population of 384,000 virtual individuals, some with virtual hypertension, to provide insights into the multilevel interactions between kidneys and heart to hone treatments.[43]

To re-create the extraordinary complexity of an actual circulatory system is much more difficult, however. Peter has worked with his group at UCL and Amsterdam, along with the Foundation for Research on Information Technologies in Society (IT'IS), an independent, non-profit research foundation in Zurich, and others to re-create *in silico* a 60,000-mile-long network of vessels, arteries, veins and capillaries using billions of data points from a digitized human built from high-resolution (0.1 × 0.1 × 0.2 millimetre) colour cross-sections of a frozen cadaver of a 26-year-old Korean woman, Yoon-sun.

To simulate just her blood circulation—and nothing else—was a monumental task: Peter's team at UCL, notably Jon McCullough, had to work with the Jülich Supercomputing Centre, University of Tennessee, Leibniz Supercomputing Centre, IT'IS Foundation, Ruhr University and the University of Amsterdam to write around 200,000 lines of code that could run on 311,000 cores on the SuperMUC-NG to simulate blood flow at 434,579,134 sites within the tortuous combination of arteries and veins within Yoon-sun's body.

Peter's team had by then developed HemeLB, a highly scalable supercomputing class code that simulates the blood flow through a stent inserted in a patient's brain to treat aneurysms, bulges caused by weaknesses in the wall of a blood vessel. The software handles 3D units—voxels—of blood and simulates its flow within a given geometry. To run it, they once again turned to the Super-MUC-NG supercomputer.[44–46]

To adapt the supercomputer to simulate Yoon-sun's circulation was a long way from simply clicking on an icon. HemeLB has had to be repeatedly optimised and tuned, building in the latest standards for parallel computing and approaches to balancing the load on the machine's hundreds of thousands of computing units, cores. Only then could Super-MUC-NG run the code efficiently at a vast scale. Indeed, Peter's team has most recently developed an accelerated version of the code, which scales to 20,000 GPUs on Summit.[47]

Fortunately, as Noble had found with the Connection Machine decades earlier, the job was made easier because the biology mapped onto the supercomputer's architecture. The overall flow of blood through an artery was boiled down to a lattice of smaller blocks, and the calculations could work out the trickles of blood between adjacent blocks. This nearest neighbour-to-neighbour communication between blood fluid sites mapped neatly onto the

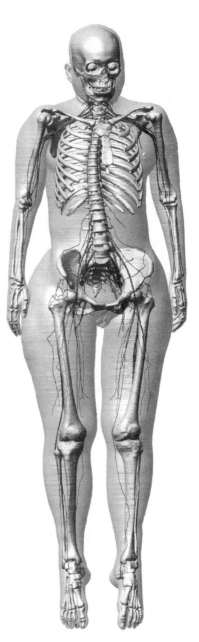

FIGURE 37. Yoon-sun. The image quality and resolution (0.1× 0.1× 0.2 mm) made it possible to reveal details of the peripheral nerves, arteries, veins and other small structures. (IT'IS Foundation)

architecture of the machine's processors and the flow of information between them.[48] (By comparison, when simulating the brain, where there are many long-range interactions, it would be daunting to map them to this particular computer architecture.[49])

The team was able to create a virtual copy of Yoon-sun's blood vessels down to a fraction of a millimetre across, though not down to the finer network of gossamer vessels that lie beyond the tips of around 1500 blood vessels, which were tackled with "subgrid" models. The virtual blood itself was relatively simple to create: the equations of fluid mechanics ignore how blood is made up of white blood cells, red blood cells, plasma and so on, treating it as a continuum. By taking over SuperMUC-NG for several days, they could show in a realistic way how blood flowed for around 100 seconds through the virtual body of Yoon-sun.

The team is now using this understanding to chart detailed variations in blood pressure throughout the body to investigate how these differences correlate with various types of disease. Another use of this circulatory model is to simulate for the first time the movement of blood clots through the body; and, most recently, to demonstrate how the inclusion of the entire human forearm in simulations of the arterial and venous trees permits accurate modelling of an arteriovenous fistula (AVF), which surgeons create by connecting a patient's own vein and artery to form a long-lasting site through which blood can be removed and returned for kidney dialysis.[50] Unfortunately, a significant fraction of fistulas fail to mature for successful dialysis use, and the hope is that this model can help improve the chance of success.

In the United States, another effort to model the circulatory system of an entire person is under way by Amanda Randles at Duke University in North Carolina to shed light on diseases such as hypertension. The fluid dynamics code is called HARVEY, after William Harvey (1578–1657), the English surgeon who first described the circulatory system. "The potential impact of blood flow simulations on the diagnosis and treatment of patients suffering from vascular disease is tremendous," she wrote with colleagues.[51] As just one example, HARVEY has helped Randles understand the treatment of cerebral aneurysms with stents.

Meanwhile, in Graz, Austria, Gernot Plank has worked with colleagues in the Netherlands and with Edward Vigmond in France to create the first three-dimensional "closed loop" model of the heart and circulatory system,[52] which, as he told Peter, "is as close to a universal electromechanical heart simulator as anybody has got so far."* These projects represent a remarkable milestone because, at last, entire human simulations as well as very sizable fractions of the human body are swimming into view.

The Modern Prometheus

There's even a hint of multiscale modelling in Mary Shelley's *Frankenstein; or, the Modern Prometheus*. In her Gothic classic, Victor Frankenstein builds his creature in his laboratory through a mysterious blend of chemistry and alchemy, along with a mix of animal and human organs. As well as hinting at the complexity of creating a human, the work ranks among the first fictional presentations of the education of a young scientist.

Some of Shelley's inspiration came from lectures delivered at London's Royal Institution by the great chemist Humphry Davy (1778–1829).[53] Given the scientific influences on the writing of *Frankenstein*, Shelley would be fascinated to know how far science has come in stitching together different kinds of descriptions to create a beating heart. To her, the heart was particularly significant, given how she behaved in the wake of the death in July 1822 of her husband, Percy Bysshe Shelley, who was only 29 when he drowned after his boat was caught in a storm.

The poet's body was cremated in the presence of his friends, who retrieved the remains of Shelley's heart from the pyre and gave them to his mourning wife. In 1852, a year after Mary Shelley died, Percy's heart was found in her desk, wrapped in the pages of *Adonais*, one of his last poems. The elegy, written to mourn the passing of John Keats decades earlier, remarks "when lofty thought, Lifts a young heart above its mortal lair."

---

* Gernot Plank, email to Peter Coveney, October 4, 2021.

The heart is not the only organ being animated with lofty thoughts, in the form of theory, along with vast quantities of patient data. Efforts to simulate a digital alter ego have extended far beyond the circulatory system to mirroring the workings of all the major organs in the human body. As we will see in the next chapter, the era of the digital human twin is fast approaching.

8

The Virtual Body

Our "Physick" and "Anatomy" have embraced such infinite varieties of being, have laid open such new worlds in time and space, have grappled, not unsuccessfully, with such complex problems, that the eyes of Vesalius and of Harvey might be dazzled by the sight of the tree that has grown out of their grain of mustard seed.

—Thomas Henry Huxley*

Virtual human cells, tissues and organ systems are being developed worldwide. There are *in silico* versions of the complex traffic of activity across genetic regulatory networks, the passage of a breath through airways, the rhythmic expansion of the lungs, along with surges of blood through a liver. There are even nascent versions of a virtual brain, without doubt the most challenging feature of a virtual human.

All these developments have arisen from taking four basic steps: gathering data about the body, developing theory, judiciously using AI and then working out how to combine and blend theories, from those that model the ways molecules interact to cells, organs and our physiology, which depends on all kinds of physics—mechanical, electrical, fluid flow, heat transfer and so on—that act over multiple scales. The fifth and final step is to animate these multiscale and multiphysics models in computer simulations so that we can predict how Virtual You behaves in different circumstances.

One sign of how far this research has come is that, as we mentioned earlier, regulators are turning to digital twins. The US Food and Drug Administration used computer-simulated imaging of 2986

* St. Martin's Hall, "On the Advisableness of Improving Natural Knowledge," *Fortnightly Review*, vol. 3, 62 (1866).

in silico patients to compare digital mammography and digital breast tomosynthesis (mammography that uses a low-dose X-ray system and computer reconstructions to create 3D images of the breasts). The increased performance for tomosynthesis was consistent with results from a comparative trial using patients and radiologists. The study's findings indicate that *in silico* imaging trials and imaging system computer simulation tools are not only useful but cheaper and faster than doing them on people.[1] Others are creating digital twins of patients—"iPhantoms"—to work out patient-specific radiation doses to their organs during imaging.*

There are many more examples, not least in the previous chapter. It is impossible to do justice to them all, but, of the multitude of teams working on the virtual human, one of the most sustained and diverse efforts can be found in New Zealand at the Auckland Bioengineering Institute (ABI) an imposing tower of dark glass in the north part of Auckland a few blocks from the waterfront. The institute is run by Peter Hunter, who spent his boyhood fixing radios and playing with electronics (his father constructed New Zealand's first closed-circuit television station in his backyard), and did his early work in Oxford with Denis Noble on the heart.

Hunter, who started out as an engineer with an interest in mathematical modelling, has been preoccupied for decades with the problem of how to simulate the body in a computer. In the introduction, we mentioned the various virtual human endeavours worldwide, and in 1997 Hunter's institute led one such effort, the Physiome Project, where *physio* denotes "physiology" and *ome* means "as a whole."[2] †

This worldwide undertaking dates back to a meeting in Glasgow, Scotland, in 1993, and was spurred on by the amount and diversity of medical data available even then. Today, we can draw on a much broader range of data, along with various atlases, from muscle regeneration to metabolism, the latter providing a valuable way to integrate the torrent of -omics data.[3,4] Fresh insights into structure and anatomy come from those who have donated their bodies, such as Yoon-sun, mentioned in the last chapter, to various visible human projects in which a cadaver is frozen, sliced tens of thousands of

---

* Wanyi Fu, "i-Phantom Framework," CompBioMed Conference, September 17, 2021.
† Peter Hunter, interview with Peter Coveney and Roger Highfield, October 23, 2020.

times—one paper-thin section at a time—and photographed after each cut to be digitized and then reassembled into a virtual person.[5]

Just as the submarine worlds of the deep oceans still harbour many secrets, there are still surprises to be found in the human body. In 2018, one team at New York University found that it contains a "highway" of moving fluid. Layers that had been thought to be dense, connective tissues—which lie below the skin's surface, lining the digestive tract, lungs and urinary systems, and surrounding arteries, veins and the fascia between muscles—turned out to be interconnected, fluid-filled compartments.[6] Perhaps they act as shock absorbers, or are a source of lymph, a fluid that is involved in immunity.

The work serves as yet another reminder that new techniques drive new understanding. This fluid highway had been overlooked by traditional methods for studying tissue under the microscope, in which the tissue was treated with chemicals, sliced thinly, and dyed to highlight key features. Fluid was drained during this fixing process, so the connective protein meshwork around once fluid-filled compartments pancaked, appearing solid. The structures were eventually revealed by using a new method, known as probe-based confocal laser endomicroscopy.

To make sense of all these insights, data and models, around 300 people work in the Auckland Bioengineering Institute, of whom 100 are pursuing doctorates, 60 are full-time researchers, 20 are academics and the rest are support staff. To support this effort requires more than government funding, and Hunter has fostered a supportive culture of entrepreneurship, encouraging researchers to set up companies worth millions of dollars that in turn employ hundreds more people. The spin-outs range from those that specialise in medical technology, for instance, who use modelling to interpret data from medical devices, to Soul Machines, a company set up by the Institute's Mark Sagar to create realistic-looking "avatars," which have appeared in showrooms, banks and Hollywood.

The Twelve Labours

Efforts are under way around the world to extend and integrate the range of virtual organs and organ systems, and to develop modelling, from simple, one-dimensional descriptions based on ordinary differential equations that can be run on a laptop to those that simulate organs in three dimensions and time, which depend on partial differential equations that can only be explored with the help of supercomputers.

Here, too, the research in New Zealand provides a flavour of the challenges. By the time we caught up with Peter Hunter, his team was extending their research from the heart to the body's 12 organ systems, including the brain. He refers to this ambition as "the 12 labours," a nod towards a penance carried out by Heracles (or Hercules), the greatest of the Greek heroes, to serve King Eurystheus, in return for which Heracles would be rewarded with immortality.

One of the hurdles that Hunter's institute faces is building the virtual infrastructure that coordinates activity across the body, such as hormones, the chemical couriers that carry messages hither and thither, and highlighted by the Guyton whole-body circulatory regulation model in the last chapter. A team led by his colleague Vinod Suresh is trying to connect organ physiology to underlying cellular processes, such as the uptake and release of chemical substances by cells, their transport through blood and sensing by tissue, which underlies a number of regulatory processes—insulin secretion in response to a meal, breathing in response to exercise and so on.

By blending medical imaging, mathematical modelling, laboratory experimentation and computer simulations, they hope to understand how the traffic of regulatory chemicals and signalling molecules affects the absorption of nutrients, the secretion of saliva and the balance of water and salt in the lungs. As for where this effort may be leading, they envisage the development of new tests for dementia, for example, since there is evidence that changes in blood flow and water transport in the brain can precede symptoms.

Another example of the body's infrastructure comes in the form of the nerves that permeate every organ, allowing them to communicate with each other and "talk" to the brain and spinal cord. The human body is wired to move, respond, sense and more, so that hundreds of activities are controlled at the same time. This housekeeping activity, which is so crucial for the everyday business of life, depends on the autonomic nervous system, consisting of the

Cell models   **Organs**   **Organ systems**   **Whole body physiology**

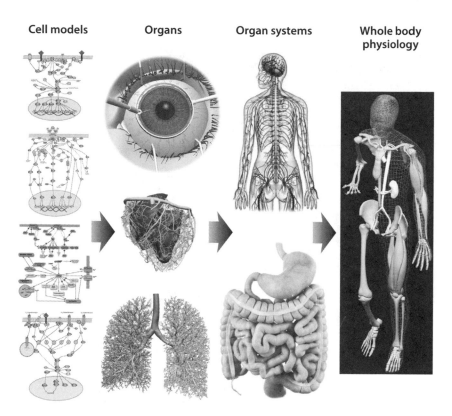

FIGURE 38. From modelling cells to the virtual body. (Peter Hunter, Auckland Bioengineering Institute)

parasympathetic "rest and digest" part and the sympathetic "fight or flight" part.

Twelve pairs of cranial nerves fan out from the underside of the brain to control head and neck muscles, or carry messages from sense organs. Among them is the vagus nerve, a nerve superhighway that snakes its way from the head through the abdomen to link the brain with major organs. The vagus has numerous roles, including in inflammatory response and appetite, and is one focus of interest for researchers developing "electroceuticals"—electrical signals—to trick the body and brain, for instance, into thinking that the gut is full.[7]

Doctors have also known for some time that nerves fine-tune a heartbeat, the rate of breathing or lower blood pressure. In the spleen, they can alter the activity of T cells—the immune cells we encountered earlier—to halt the production of inflammatory substances, such as tumour necrosis factor, a protein that accumulates in joints in rheumatoid arthritis. And there are nerves in the skin that can suppress infection. Rather than expose the entire body to a drug that circulates in the bloodstream, an electroceutical implant could tweak activation of a single nerve, such as the one that controls the bladder.

To aid the development of electroceuticals, the Auckland team is generating maps of the nerve pathways that connect organs, from the ulnar nerve that bends your little finger to the common peroneal nerve that lifts your foot. We need deeper understanding of the wiring that couples the heart and lungs, for example, because, as Hunter points out, the heart is full of neurons, and "when you breathe in, your heart beats slightly faster—that makes sense because you want blood going through the lungs when oxygen levels are high."* That clever feat is handled by the body's autonomic system, via clusters of nerves, called ganglia, along with the brain stem.

Researchers in Auckland are developing tools to influence the neural circuitry that interconnects a wide range of tissues. Known as SPARC (stimulating peripheral activity to relieve conditions), a program backed by the US National Institutes of Health, they hope these tools will help develop new treatments for diverse conditions, such as gastrointestinal disorders, hypertension, heart failure, type 2 diabetes, inflammatory disorders and more.[8]

Even heart attacks can be caused by what Hunter calls an "imbalance" in the interlocked feedback control loops of the body's autonomic nervous system, and not just by heart muscle starved of blood by coronary disease. We already implant pacemakers to regulate the activity of the heart, and the hope is that implantable devices could take over from an ailing autonomic nervous system to help regulate the heart, gut and more.

To model these regulatory feedback loops and daisy chains of dependencies between brain stem, ganglia and nerves, the Auckland team uses coupled ordinary differential equations. However, as we have already seen, when it comes to the heart and lungs, for example, they also have to use a spatial model, where partial differential equa-

---

* Peter Hunter, interview with Peter Coveney and Roger Highfield, October 23, 2020.

tions model the differences in local anatomy, whether the chambers of the heart or the way oxygen transfer in lungs is dependent on gravity, leading to varying amounts of oxygen and carbon dioxide exchange at different levels in the lung.

Getting hold of all the parameters to set up complex multiscale models is a major hurdle. Here the Auckland team, typical of many other groups, is using surrogate modelling that we encountered earlier. The team can substitute the full physics of a highly detailed multiscale model with machine learning that is trained to behave the same way, so that complex equations do not have to be solved time and time again. A key ingredient of success, Hunter reminds us, is to ensure machine learning is constrained by real physics—such as laws of conservation of energy—to ensure that it does not produce, as he says, "silly answers when it hits a situation that is sufficiently different from its training data set. That is when you want the physics to kick in."*

Virtual Muscles and Skeletons

Another key piece of human infrastructure is the body's supporting muscles, bones, cartilage, ligaments, tendons and other connective tissues. Here simulations can help make sense of an epidemic of musculoskeletal and movement disorders, a global health issue linked with our increasingly aging and overweight population. Computer simulations can help explore the interplay between micro and macro scale in human joints during growth, aging, and disease.

We have already encountered some of the work done in Europe. Some, shown in our *Virtual Humans* movie, was carried out by Marco Viceconti at the Universities of Bologna and Sheffield, respectively, and Alberto Marzo and their colleagues at the University of Sheffield, who used finite element analysis to show how to predict the risk of bone fracture in elderly patients with osteoporosis, when aging bones lose their ability to regenerate, so they shed mass, change structure and become brittle. The team has also studied the effects

* Peter Hunter, interview with Peter Coveney and Roger Highfield, October 23, 2020.

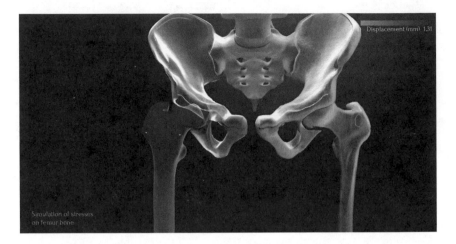

Displacement (mm) 1.31

Simulation of stresses
on femur bone

FIGURE 39. Simulation of stresses on a human femur. (CompBioMed and Barcelona Super-computing Centre)

on bone of metastases, when cancer spreads from a site such as the prostate to the spine to cause "lytic metastases," which destroy bone tissue.[9] In yet another project, they modelled spines that had been twisted by scoliosis to understand how the curvature changes the stresses and strains.[10] These computational models can also help doctors plan surgery to improve gait in children affected by cerebral palsy,[11] or treat craniosynostosis, a birth defect in which the bony plates of an infant's skull prematurely fuse.[12]

With US regulators, the Auckland team is working on a Musculo-skeletal Atlas Project, a software framework that aims to streamline the regulation of orthopaedic devices with virtual tests. Companies such as Ossis Ltd., based in New Zealand, are accelerating the manufacture of customised 3D-printed titanium hip implants, while the Australian Institute of Sport is monitoring athletic performance via wearable wireless sensors.

The Auckland team has developed computational models informed by novel sensing and medical imaging to explore various practical issues: Can "smart splints" help children with neurological disorders? Can we reduce the burden of osteoarthritis by changing how we move? Can we improve recovery from stroke with robotic rehabilitation? More fundamental anatomical issues are being tackled

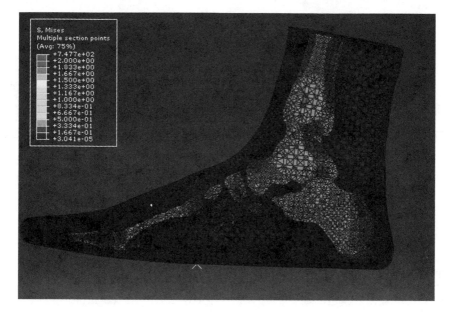

S, Mises
Multiple section points
(Avg: 75%)
+7.477e+02
+2.000e+00
+1.833e+00
+1.667e+00
+1.500e+00
+1.333e+00
+1.167e+00
+1.000e+00
+8.334e-01
+6.667e-01
+5.000e-01
+3.334e-01
+1.667e-01
+3.041e-05

FIGURE 40. Digital orthopaedics. (Reproduced with permission by Dassault Systèmes)

by Thor Besier, who is using digital twins to explore the role of the thumb in the primate world, where thumbs offer different degrees of dexterity depending on whether they are used to grasp, climb, hang, peel or throw.

In France, Dassault is also developing a virtual musculoskeletal system for knee and hip implants, and to study problems with feet, shoulders and spines. When a patient has an abnormal walking test, for example, the model can provide insights into what is going on at the muscular level. To plan surgery on the complex anatomy of the foot and ankle, which comprises 28 bones, 40 joints, more than 100 ligaments and 30 tendons, the Dassault technology is used by a Belgium-based start-up called Digital Orthopaedics.

Virtual Lungs

Every breath you take can now be simulated in a computer. The lungs contain a vast branching network of passages, which end in millions of air sacs through which the oxygen from an inhalation is exchanged for carbon dioxide, and are so rich in blood vessels they are pink. They contain dozens of cell types, each with specialised roles, such as delivering oxygen and eliminating carbon dioxide from blood, or secreting surfactant and mucus to lubricate air spaces for expansion and contraction, or creating a barrier to pathogens and harmful pollution.

One complication is that the lungs are lopsided, so both have to be individually modelled. Your right lung is shorter than your left, to accommodate the liver, and your left lung is slightly shrunken to accommodate your heart. Each lung is separated into sections called lobes, and the right lung has three and the left only two. The lungs also present a multiscale simulation because they span several orders of magnitude, from the separate lobes to the airways and blood vessels that can be as big as 25mm and bifurcate into ever smaller branches, eventually terminating in capillaries only 0.008 mm across.

Ultimately, a lung's ability to work relies on its functioning like a bellows. Apprehending this pumping action helps us understand major respiratory diseases, such as asthma and chronic obstructive pulmonary disease, a progressive, complex mix of airway inflammatory disease, chronic bronchitis and emphysema that is often linked to smoking.

As ever, there has been an explosion of data at the molecular and genetic level, along with ever more detailed imaging of the whole lung. It is possible to take high-resolution scans to model its lobes and geometry as a mesh. Individual cells can be modelled too. However, as Merryn Tawhai, deputy director of the Auckland Bioengineering Institute, Kelly Burrowes and their colleagues point out, relatively little has been done to link events at these extreme scales. The extent to which this has to be done in virtual versions depends on the aim of the simulation: simple one-dimensional models are sufficient to show how heat and humidity move through the organs,[13] while complex three-dimensional models are needed to capture the details of turbulent airflow as particles of a drug are inhaled.[14]

By drawing on medical imaging data, Merryn Tawhai's team in Auckland creates personalised patient-based and species-specific geometric lung models.[15] Their virtual model captures the geometry of the organ and includes the lung's lobes, the 30,000 airways that transport oxygen to and carbon dioxide from, as exchange units deep in the lung, and the pulmonary circulation through arteries, veins and capillaries, along with subcellular processes, notably gas exchange. The virtual lungs show, for example, that there is a clear relationship between lung shape and function, revealing the influence of age, as their capacity declines, the diaphragm weakens and alveoli become baggy.

With the help of a computational model, the team can simulate breathing and link resistance to airflow within the conducting airways to the deformation of the alveolar/acinar lung tissue in the lung's branched airways, where an impressive amount of breathing apparatus is squeezed into a relatively modest space. You can even inspect the lungs online and see how, when taking a gulp of air, the tissue closest to the diaphragm (the muscle that sits under the lungs) expands the most, while the tissue atop the lung expands the least.[16]

The virtual lungs reveal the impact of asthma, which causes airways to shrink as the smooth muscle that wraps around them constricts and the airways secrete mucus, demanding more energy to take each breath. That is important given we take around 550 million inhalations during a lifetime. Another simulation shows the effects of smoking, which damages the airways and gas exchange tissue, making it harder to get enough oxygen, while the lung tissue breaks down and becomes floppy. The team has also used the model to understand lung changes in patients with acute pulmonary embolism, a blockage of a pulmonary (lung) artery, typically caused by a blood clot.[17]

In related work, Jim Wild's group at the University of Sheffield is using imaging and modelling to provide deeper insight into pathological changes in the lung to develop novel diagnostic techniques,[18] while a virtual respiratory tract is under development at the US Department of Energy's Pacific Northwest National Laboratory in Richland, Washington. Their aim is to give an unprecedented, three-dimensional view of how pollutants enter, traverse and collect in the respiratory system. Having modelled the nose, larynx and lungs of a rat, the team has extended the work to humans to understand the impacts of poor air quality,[19] which has become more important, as evidenced by the nearly doubling of asthma sufferers since 1980. The long-term health effects from air pollution include heart disease, lung cancer and respiratory diseases.

By learning how particulates travel through the lungs, scientists can design treatments that more precisely target drug delivery for pulmonary diseases. Our *Virtual Humans* film featured another project by Denis Doorly at Imperial College London, Mariano Vázquez of the Barcelona Supercomputing Center and colleagues, which re-

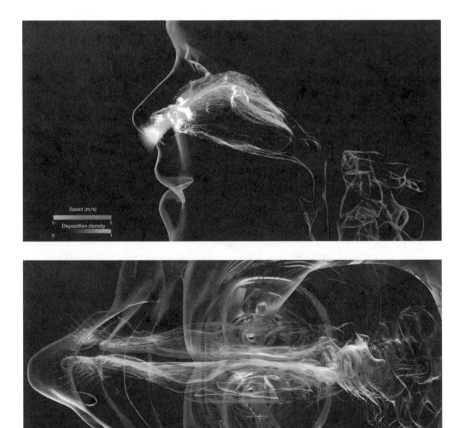

FIGURE 41. A virtual breath. (Credit: CompBioMed and Barcelona Supercomputing Centre)

vealed how, when we take a sniff, air drawn through our nose undergoes successive accelerations and decelerations as it is turned, split and recombined before dividing once again at the end of the trachea as it enters the bronchi that deliver a breath to each lung.[20]

They are also able to study how pollutants impact the lungs of healthy people compared with those who suffer from respiratory ailments. For example, they can begin to simulate how gases, vapours and particulates may act differently within lungs of people suffering from cystic fibrosis, emphysema and asthma. Within a decade, there will be full-blown models of human respiration that can be customised to speed diagnosis and treatment for an individual patient.

Virtual Liver

The liver is a remarkable organ, the largest within the body—second in size only to your skin—and the busiest one too, acting as a processing plant, a battery, a filter, a warehouse and a distribution centre. The liver does many critical jobs, transforming food into energy, detoxifying alcohol, drugs and pollutants, storing vitamins, fat, sugars and minerals, manufacturing a digestive juice—bile—which breaks down fats so they can be absorbed by the body and producing hormones that regulate sexual desire and more. It is the chief metabolic organ of all vertebrates, their central laboratory if you like, and its metabolism is central to drug toxicity and efficacy.

Roughly cone-shaped, the liver consists of two main lobes. Each one has eight segments that consist of 1000 small lobes, or lobules, connected to ducts that transport bile to the gallbladder and small intestine. Because these lobules are the basic functional units of the liver, an Auckland team led by Harvey Ho has developed a multiscale virtual lobule model in which flow is simulated with a partial differential equation accounting for the diffusion and advection of chemical species to link biomechanical models of blood flow and pharmacological models. In this way, they can, for example, see how in one particular metabolic zone in the lobule an overdose of the painkiller acetaminophen may lead to the necrosis of hepatocytes—the most common type of cell in the liver.[21]

Unlike any other organ in the body, the liver relies on two blood supplies: one-quarter comes via the hepatic artery, which delivers highly oxygenated blood from the lungs. Three-quarters of its blood arrives through the portal vein system, a network that transports blood through the intestine, the stomach, the spleen and the pancreas. All the products of digestion, from nutrients to toxins, pass into the liver via this route. Once the liver has deoxygenated and processed this blood, it is transported to the liver's central hepatic vein and then to the heart.

As the most important metabolic organ in your body, your liver is so active that it warms the blood passing through it, which helps maintain the body's temperature. In fact, the liver is saturated with blood, holding something like a quarter of your body's supply, which is why this great organ is reddish brown in hue. In all, every two and a half minutes a gallon of blood passes through the liver's complicated network of arteries, veins and capillaries. One of the efforts in Auckland aims to map out these great trees of thousands of vessels,[22] along with the ebbs and flows of blood that pulse through them, to plan liver operations.[23] The liver has few external landmarks to help guide surgeons, so models of its anatomy can help with replumbing after major surgery.[24]

The liver is also unusual because, even if more than half of its overall mass is damaged—for instance, by drinking too much alcohol—it can regenerate itself, a process that is also being simulated. The work has already revealed an unrecognised mechanism of regeneration, which depends on how hepatocytes align themselves along the sinusoids, the micro blood vessels that traverse the liver lobule.[25] In another study, the Auckland team is modelling the effects of a hepatitis B viral infection on the liver, using the understanding of viral kinetics that we outlined in Chapter Five.

Other hepatic models thrive, such as the Virtual Liver Network in Germany, where a truly multiscale computer model of the complete organ—from the biomolecular and biochemical processes within cells up to the anatomy of the whole liver—is under development by 70 research groups from 41 institutions. For example, Marino Zerial and colleagues at the Max Planck Institute of Molecular Cell Biology and Genetics in Dresden have developed a multiscale model that can simulate the fluid dynamics properties of bile, which is produced in the liver and transported through a highly branched tubular network to the intestine to digest lipids and excrete waste products.[26] Their 3D representations of the biliary network, a great tangle of vessels, show the fluid dynamics of bile, from fast, flowing red to sluggish blue. This kind of work helps us understand cholestasis, the impairment of bile flow, a common problem seen in drug development and that can be linked to a drug overdose, among other things.

When it comes to the pancreas, which produces juices that aid in digestion, artificial versions have been developed for type 1 diabetes patients, who lack insulin or are unable to respond to it adequately. To help them get the right amount of insulin to regulate their blood glucose levels, the artificial variety uses a mathematical model of human glucose metabolism and a closed-loop control algorithm that models insulin delivery using data from a glucose sensor implanted in the patient. These can be customised into a patient-specific digital twin of the pancreas that continuously calculates how much insulin is required and controls an implanted pump to maintain blood insulin concentrations.[27]

Virtual Gut

To help understand the breakdown of food in the body, an Auckland team led by Leo Cheng has developed mathematical models of the stomach and small intestine—a virtual gut. Overall, they have found ways to overcome multiscale challenges in modelling the electrophysiological processes from cellular to organ level occurring during digestion.

Much like the heart, the gut can suffer from arrhythmias. At the cellular level, models are now being applied in various ways, from investigations of gastric electrical pacemaker mechanisms to measuring the effects of cirrhosis on gastrointestinal electrical activation. At the organ level, high-resolution electrical mapping and modelling studies are being combined to provide deeper insights into normal and dysrhythmic electrical activation.

These electrifying journeys through the gut are also enabling detailed modelling studies at the "whole-body" level, with implications for diagnostic techniques for gastric dysrhythmias and paralysis, when food does not pass through the stomach, and how to treat them with pacemakers or ablation, when a little tissue is killed off to help restore the gut's natural rhythm.

The team also wants to take into account a major shift in understanding of the human body that occurred a decade or two ago, when interest moved from human cells to the bacteria and other microbial cells in the gut. It used to be thought that these passengers outnumber our own cells by 10 to 1. We now know that this ratio varies, but it has been estimated that "reference man" (who is 70 kilograms, 20–30 years old and 1.7 metres tall) contains on average about 30 trillion human cells and 39 trillion bacteria, such that each defecation may flip the ratio to favour human cells over bacteria, as the latter falls by a quarter to a third.[28]

Several thousand strains of bacteria graze in the human gut. Some are associated with disease, while others have beneficial effects. Surprisingly, despite our close genetic relationship to apes, the human gut microbiome is more similar to that of Old World monkeys like baboons than to that of apes like chimpanzees.[29] Even more surprising, the microbial contents of the gut can affect the brain, perhaps even memory.[30]

The Auckland team is using laboratory experiments, growing populations of the bacteria in a bioreactor, carrying out metabolomics and bioinformatic analyses and developing mathematical along with statistical techniques to craft predictive models that will improve our ability to understand and manipulate these microbiomes. To help make sense of the remarkable complexity of the microbial flora and fauna, the researchers are resorting to using machine learning to find out, when nudged with a change in temperature, food or whatever, how microbial populations change in the gut.

Virtual Metabolism

To date, scientists have successfully simulated drug action, infections, cell tissues, organs and more. As mentioned earlier, efforts are now under way to integrate these aspects of the virtual human with a virtual metabolism—that underlying network of genes, proteins and biochemistry that straddles the human body.

Some models are simple. To probe the complex mechanisms underlying muscle fatigue, Yukiko Himeno and colleagues at Ritsumeikan University in Kyoto developed a mathematical model with five basic components—muscle, liver, lungs, blood vessels and other organs—to caricature the key mechanisms. The simulation results provide the essential evidence for a better understanding of muscle fatigue, such as increases in blood lactate and muscle inorganic phosphate, along with the fall in blood pH level, that results from high-intensity exercise.[31]

The first computer model to represent the three-dimensional structural data of proteins and metabolites in metabolic processes was developed by an international consortium including Bernhard Palsson at the University of California, San Diego, and Ines Thiele, who now works at the National University of Ireland in Galway. The model covered around 4000 metabolic products, or metabolites as they are known, and nearly 13,000 protein structures, all united within a computer-based tool called Recon3D, which gives insights into genetic variation and the mechanisms underlying the effects of drugs on metabolic response in humans.[32]

Using Recon3D, they can, for example, study in detail how metabolic processes run differently in Parkinson's patients compared with healthy people, study the interplay between pathogens, such as bacteria or viruses, and their human host or investigate the effects of a genetic mutation on the structure of a metabolite or protein during the development of certain diseases. They also used Recon3D to study how genes, proteins and metabolic reactions react to various treatments. To their surprise, the researchers found that drugs with very different molecular structures can produce similar metabolic responses.

Based on this work, Thiele and her colleagues developed a new metabolic network reconstruction approach that used organ-specific information from the literature—a survey of more than 2000 articles and books—and -omics data to capture the metabolism of 26 organs and six blood cell types in two individuals, one male and one female, Harvey and Harvetta.[33]

They reported in 2020 how each of their reconstructions represents the metabolism of the whole body with over 80,000 biochemical reactions in an anatomically and physiologically consistent manner, priming their reconstructions with physiological, dietary and metabolomic data to reproduce interorgan metabolic cycles and energy use, along with biomarkers of inherited metabolic diseases in different biofluids, and basal metabolic rates. Finally, by adding microbiome data, they were able to explore host-microbiome cometabolism. They concluded that the work represents "an important step toward virtual physiological humans."

Virtual Brains

The 3 pounds of tissue between your ears deliver vast computational power, with estimates ranging up to 10 to the power of 28 flops. This far exceeds the 10 to the power of 18 flops of an exascale machine, though implicit in this claim is the idea that the brain really is something like a digital computer. One sign that it is not is that your analogue grey matter only requires a modest 10 to 20 watts of power to think. While the Frontier exascale supercomputer has to be cooled by thousands of gallons of water per minute, the human brain runs at just 38.5°C.[34] To reduce the brain to flops feels as meaningless as weighing up the significance of a work of art by its size.

The brain is without doubt the most complex and the least well understood within the human body. Novel insights continue to emerge as we examine it more closely, for instance, from a recent study of a sesame seed–sized portion of brain tissue from the cerebral cortex of a 45-year-old woman who had epilepsy surgery at Massachusetts General Hospital, Boston.[35] After being stained with heavy metals, sliced into 5000 pieces and studied with electron microscopy, a team from Harvard University, Google and elsewhere used computational methods to render the three-dimensional structure of 50,000 cells, hundreds of millions of projections from neurons and 130 million synaptic connections. The study revealed a few cells that connected 10 or even 20 times. Given a similar observation in mice, perhaps sparse powerful inputs may be a general feature of mammalian brains.

This 1-millimetre cubed piece of brain contained 20,139 oligo-dendrocytes, a kind of glial cell, and revealed that glia outnumber neurons by 2 to 1 (32,315 versus 16,087) along with 8096 vasculature cells, 10,531 spiny neurons, of which 8803 had a clear pyramidal shape. The spiny cells accounted for 69% of the neurons and the 31% nonspiny neurons were classified as interneurons (4688). There was a subset of another 868 neurons that did not easily fit this binary categorisation. Given the complexity of even this little piece of brain, it will be a monumental task to simulate all its 86 billion brain cells (neurons), each of which has an average of 7000 connections to other neurons (synapses).

Even so, remarkable progress has been made. Scientists around the world have created maps that catalogue and map the diverse cells in the brain and their properties.[36] Drawing on these and other data, silicon brains and brain simulations are among the many objectives of the EU's Human Brain Project, which draws on the efforts of more than 700 scientists at over 140 universities, hospitals and research centres across Europe, along with five supercomputing centres, in-cluding the Barcelona Supercomputing Center.

Some aspects of this suite of research programmes aim to get basic insights, such as the discovery that high-frequency oscillations convey "information packages" from one small neuronal group to another group a long way away (from the cellular perspective) in the brain,[37] drawing a whole-brain map of the connectivity of the human cerebral cortex,[38] or developing a mathematical framework to quanti-tatively reproduce the entire range of a brain network's connectivity.[39] Others are tackling more practical problems, such as ways to share relevant medical brain data held by thousands of hospitals.

Modelling is a key ingredient. Some teams are developing simpli-fied yet faithful models of dendrites, the intricate treelike structures that stretch between neurons, which do so much more than simply collect signals: dendrites integrate and compare input signals to find special combinations that are important for the neurons' role.[40] Others have used computer simulations of the spinal neuronal cir-cuitry to build on research that has enabled paraplegics to walk again through epidural electrical stimulation to use targeted spinal cord stimulation to allow patients with spinal cord injuries to regain control of their blood pressure.[41] Another EU initiative, Neurotwin, aims to develop brain twins that can predict the effect of transcranial electromagnetic stimulation, or TMS,[42] when magnetic fields are used to cause currents to flow in specific regions of the brain, for instance, to treat depression, and for basic research—as one example of the power of this approach, Roger took part in an experiment to use TMS to temporarily turn off his ability to speak.[43]

Another key ingredient of the Human Brain Project is The Virtual Brain, TVB, an open-source simulation platform that blends experimental brain data from a range of sources to improve understanding of the brain's underlying mechanisms. The brain network simulation engine, based on neuronal population models and structural information from neuroimaging, has been under development since 2010 led by Viktor Jirsa, director of Inserm's Institut de Neurosciences des Systèmes at Aix-Marseille University, and collaborators Randy McIntosh at Baycrest Center Toronto and Petra Ritter at the Charité hospital in Berlin.

These virtual brain models are capable of simulating brain imaging that is routinely carried out in hospitals. One example of how the virtual brain can be used is in treating epilepsy, which affects around 50 million people worldwide. In many cases, the seizures that mark the disease can be controlled by drugs. However, close to a third of all patients are drug resistant, and the only option is for surgeons to remove the epileptogenic zone, the epicentre, where a seizure first emerges in the brain and then spreads. This area has to be identified as precisely as possible, but it is not easy and, as a result, surgery success rates are only about 60%. To help clinicians plan this difficult surgery, the virtual brain team creates personalised brain models of patients and simulates the spread of abnormal activity during seizures.[44] The "virtual epileptic patient" (VEP) was being tested on more than 350 patients in 13 French hospitals at the time of writing, representing the first example of a personalised brain modelling approach entering the clinic.*

---

* Simona Olmi, email to Peter Coveney and Roger Highfield, April 12, 2021.

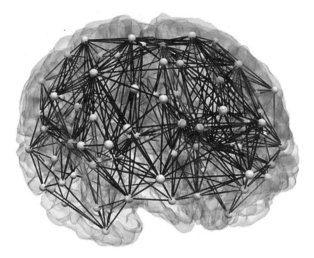

FIGURE 42. The Virtual Brain: Reconstruction of brain regions and where they are connected, shown by the nodes. (Image courtesy of the Institut de Neurosciences des Systèmes [INS], Inserm, Marseille)

First, a personalised brain model is crafted from the patient's own data on their brain anatomy, structural connectivity and activity. Through a series of steps, these data are fashioned into a dynamic network model to simulate how seizures propagate through the brain. High-performance computing and machine learning can customise brain network models to the individual patient and establish a patient-specific *in silico* brain simulation platform to test clinical hypotheses and develop treatments. At the time of writing, the EU's Human Brain Project derived its first spin-off company, Virtual Brain Technologies (VB Tech), with the ambition to move this innovation from the laboratory to patients. Jirsa told us that, using VEP as its first technology, VB Tech aims to develop personalised virtual brain technology to transform the diagnosis and care of patients suffering from brain disease.*

Personalised brain models could be used to predict the effects of tumours and help plan how to remove them with brain surgery, which has to leave as much of the surrounding tissue intact as possible for obvious reasons. Imaging methods are often used to plan an operation, but they are too crude to predict postsurgical outcomes because of the complex dynamics of the brain.

---

* Viktor Jirsa, email to Peter Coveney and Roger Highfield, October 9, 2021.

Using the open-source software of The Virtual Brain, Hannelore Aerts and a team led by Daniele Marinazzo modelled 25 individual brain networks of brain tumour patients and of 11 controls. They found these individualised models can accurately predict the effects of the tumours on brain connectivity. This result opens the possibility of using virtual brain modelling to improve the planning of this surgery and, as a consequence, its success.[45]

Virtual brain researchers are also studying strokes, a major cause of death and disability, when the blood supply to the brain is disrupted, whether by a bleed or the blockage caused by a clot. Ana Solodkin of the University of California, Irvine, and her colleagues used The Virtual Brain and data on 20 individuals who had suffered a stroke, along with 11 controls, to investigate the impact on brain dynamics. These studies provide profound insights into the effects of clearly located brain lesions on the overall functionality and health of the organ. The team hopes to decipher how the complex networks of the brain react to the isolated failure of some nodes—and the best routes to recovery for individual patients.[46]

Like the many other teams who are working on brain simulations, the Auckland team is also tackling our most complex organ, one issue at a time. Their model of the body's entire cardiovascular

Blood flow through the Circle of Willis connecting arteries to the brain

FIGURE 43. Blood flow through the Circle of Willis, connecting arteries to the brain simulated using Peter's HemeLB code. (CompBioMed and Barcelona Supercomputing Centre)

system is particularly relevant because neurons are more sensitive to oxygen deprivation than other kinds of cells, which have lower rates of metabolism.

To link their work with the cellular level, the team is working with Maryann Martone of the University of California, San Diego, who leads the US Neuroscience Information Framework. Martone is interested in how multiscale brain imaging can help integrate neuroscience data on brain anatomy, diseases, molecular biology and so on that have been acquired by different researchers on different populations of patients using different techniques and different sets of data. The Framework, which was launched in 2008, provides the tools, resources and standardized ontologies to unite these disparate data.[47]

The Auckland team is studying cerebral autoregulation, a feedback mechanism that ensures the right amount of blood supplies cerebral vasculature depending on the oxygen demand by the brain. The signals of neurons and glia trigger vasodilation, the widening of blood vessels in which endothelial cells, pericytes, and smooth muscle cells act in concert to get more blood into the right place to deliver oxygen and nutrients as well as remove the waste by-products of brain metabolism. Their model can simulate normal so-called neurovascular coupling and when it is disrupted by hypertension, Alzheimer's disease and ischaemic stroke.

Another focus of interest is brain injury, which is a growing health and socioeconomic problem as doctors get better at saving the lives of patients with serious head trauma, notably as a result of car crashes. The impact of these injuries remains poorly understood, making it hard to deal with short-term and long-term risks, ranging from brain haemorrhage to dementia. Hunter hopes that multiscale computational models, wearable sensors and handheld devices to measure impact can be blended with cutting-edge neuroscience, based on culturing brain cells, to identify injury threshold and possible targets for effective treatment.

Virtual You may even have a sense of smell. Hunter's colleagues are using AI techniques to work out the plumbing and nerve pathways associated with our ability to discern odours, when information collected and processed by the nose arrives in a brain structure, called the olfactory bulb. Airborne molecules linked to scents trigger receptor cells lining the nose to send electric signals to nerve-ending bundles in the bulb called glomeruli, and then to brain cells (neurons). However, past experiences modify the way the neurons exchange the chemical information they receive from the nose, which is how the brain can organise chemically distinct but similar smells, such as orange and lemon, into a category known as citrus.[48] When it comes to brain simulations, can we go even further?

Virtual Consciousness

Just before he was about to address the 2020 Tucson Science of Consciousness Conference, and a few days before he won a Nobel Prize in Physics for his work on black holes, Roger Penrose took the time to talk to us about the potential of virtual brains. We consulted Roger while writing our previous books because of his remarkable work on general relativity, quantum mechanics and cosmology.

When it comes to Virtual You, we were interested in his ideas about consciousness. Some argue that Penrose's 1989 book, *The Emperor's New Mind: A Search for the Missing Science of Consciousness*, helped launch modern approaches to the science of consciousness, not least because it contained a provocation that, when it comes to simulating the human brain, digital computers are not the answer.

As a schoolboy, Penrose had not only wanted to follow his parents into medicine but also harboured a secret ambition: "I was going to open people's heads up and look at how their brains worked."* Though he had to give up his ambition to become a doctor to pursue his love of mathematics, his interest in the brain was rekindled in the early 1950s, when he attended lectures as a graduate student in mathematics at the University of Cambridge. Penrose realised that conscious understanding of mathematics could not be the product solely of computational processes. His epiphany came during a mathematical logic course by S.W.P. Steen that had discussed the

---

* Roger Penrose, in conversation with David Eisenbud, "An Evening with Sir Roger Penrose," Royal Society, June 8, 2022.

influential research on the limits of computation by Alan Turing,[49] Kurt Gödel and Alonzo Church[50] that we encountered in Chapter Two: "I thought there had to be some way of transcending computability, that what we seem to be able to perceive in mathematics are things that are outside computable systems."

Penrose argues that the brain, the most complex feature of Virtual You, is not a "meat computer" and that human consciousness cannot be captured in the form of algorithms. Penrose wrote: "We do not properly understand why it is that physical behaviour is mirrored so precisely within the Platonic world, nor do we have much understanding of how conscious mentality seems to arise when physical material, such as that found in wakeful healthy human brains, is organised in just the right way. Nor do we really understand how it is that consciousness, when directed towards the understanding of mathematical problems, is capable of divining mathematical truth. What does this tell us about the nature of physical reality? It tells us that we cannot properly address the question of that reality without understanding its connection with the other two realities: conscious mentality and the wonderful world of mathematics."[51]

Since he first became interested in consciousness in the 1950s, Penrose has mulled over an enduring puzzle. Because the universe is ruled by physical laws—general relativity at big length scales and quantum theory at microscopic scales—how could they translate into the seemingly noncomputational actions of conscious understanding? "I felt that there had to be something basically missing from our current physical theories." The answer lay in quantum mechanics, which we return to in Chapter Nine.*

Working with Stuart Hameroff of the University of Arizona, Penrose has argued that structures in brain neurons, known as microtubules, are the key to solving the problem: "Consciousness derives from quantum vibrations in microtubules, protein polymers inside brain neurons, which both govern neuronal and synaptic

---

* According to Roger Penrose, "There are features of physical laws . . . which go outside the computable laws that we know about, and the place I could find room for such laws is in the collapse of the wave function, a feature of quantum mechanics which is not explained in current theory."

function, and connect brain processes to self-organizing processes in the fine scale, 'proto-conscious' quantum structure of reality."[52]

They put forward this idea in the early 1990s, and it was greeted with scepticism because the brain was considered too "warm, wet, and noisy" for quantum processes. Even though Penrose and Hameroff have since gone on to cite evidence of warm quantum coherence in plant photosynthesis, bird brain navigation, our sense of smell and brain microtubules, their idea has gained little traction.

However, leaving aside the not inconsiderable problem that we lack a cast-iron definition of consciousness, Penrose's reservations about machine intelligence ring true: *consciousness is not algorithmic*. We also lack the means to emulate how the brain evolves in a plastic fashion, though generative adversarial networks (GANs) we encountered earlier—where neural networks play off each other—mark a small step in the right direction. Moreover, as we saw in Chapter Two, the limited strings of binary that inhabit digital computers probably lack the complexity to allow anything like consciousness to emerge.

Virtual You

We have come a long way since the first signals rippled through the virtual nerve cells of Hodgkin and Huxley, Turing modelled the patterns of life with mathematics and Noble persuaded a virtual heart cell to beat. Today's global effort to create digital twins is spurred on by improvements in computer hardware, advances in theory, and the never-ending refinement of modelling multiscale processes in the body, along with gathering the right data, which means medical databases that have young and old patients, women as well as men, and are diverse in every way. We also need the careful application of artificial intelligence methods.

At the superficial level, you might think the effort to create Virtual You is well advanced. There are indeed cell models, such as of a cardiac myocyte, that can simulate everything from membrane ion currents to the forces generated by filaments of protein. We also have models of small regions of tissue that have a well-defined function, from bone osteons to kidney nephrons, lung acini, lymph nodes, muscle motor units, sweat glands, cortical columns and liver lobules. Then there are models of whole organs, such as heart, lung, stomach and colon, and organ systems too, such as cardiovascular and cardiorespiratory systems—heart, vasculature and lungs—the neuromuscular system, that is, the innervation of muscles, gastrointestinal system, female reproductive system and autonomic nervous system.

Work has already started to personalise models, when anatomical data drawn from noninvasive high-resolution imaging can be used to prime algorithms that, when put to work on powerful computers, resolve the resulting equations and unknowns, generating a complex mathematical model of an organ's function, along with a virtual organ that looks and behaves like your own. There is a vast amount of effort required to provide numerical values for the many parameters to prime these webs of equations, which have to be repeated for each and every person to generate a personalised model. Here, again, judicious use of machine learning can help. All along, there is an enduring need to ensure reproducibility, and to understand the inherent uncertainties, so guaranteeing that the forecasts of a virtual human can be trusted.

While machine learning and AI models based on big data can scour past health records to see how the body will behave in similar conditions, and infer how it will change over time, a digital avatar can apply the laws of physics and mathematics to make predictions. A heart twin, for example, can model the stiffness of the heart, or the pressure gradients within the heart,[53] and simulate how the organ will behave in unusual conditions.

All this is necessary, but it is not sufficient to create Virtual You. Many of the current generation of models are based on ordinary differential equations that only work in one dimension (time). They are relatively ad hoc, general and simplified models, not personalised ones. They cannot be assembled into some kind of virtual human, because they remain bits and pieces of virtual human customised to work with isolated puddles of data with given algorithms on specific computers. Despite much work on interoperability, these fragmentary models cannot be easily connected into a personalised human. If they could, Virtual You would already be a reality. These simplified models fall far short of a nonlinear, dynamical system of the complexity of a human that involves feedback loops at all lengths and scales. That work has only started in earnest in the past decade or so.

With the rise of exascale computing, the ability to model the complexity of the body and the brain will make great strides, if the experiences with HemeLB to simulate circulation are anything to go by. The number-crunching power of pre-exascale machines, such as SuperMUC-NG and Summit, is already consumed by only a small chunk of the whole human data set. To do more, by extrapolation, we will surely need many exascale machines. However, there is no end to what we could do with more computer power at the zettascale and beyond. The same goes for the latest generation of heterogeneous supercomputer architectures that mimic more of the heterogeneous architecture of human tissue as we seek to assemble a virtual body, initially in a multiphysics fashion over the same length and times-cales, then over many scales, from genome to cell to organ. In the next chapter, in an extension of the third step we outlined in Chapter Three, we will see how multiphysics and multiscale modelling will be accelerated by the development of novel kinds of computers. Then we can expect to see more reliable and detailed models, not least a true human virtual twin.

Virtual You 2.0

On two occasions I have been asked, "Pray, Mr. Babbage, if you put into the machine wrong figures, will the right answers come out?" I am not able rightly to apprehend the kind of confusion of ideas that could provoke such a question.

—Charles Babbage, *Passages from the Life of a Philosopher* (1864)[1]

The remarkable progress so far in the creation of Virtual You is reflected in the vast range of human simulations, from molecules to organelles, cells, tissues, organs and even at the scale of the entire human body. But all this work rests on the idea that we are able to reproduce the workings of the world in a digital computer. Is this assumption correct? The bad news is that, as we saw in Chapter Two, the ability of the analogue human brain to understand Gödelian statements surpasses the capability of any digital computer. Nor can we always be sure that digital computers give the right answers. The good news is that, as well as using ensemble methods, there are ways around this hurdle by going back to the future of computing.

As we saw with the Antikythera mechanism, the first computers were analogue. Only relatively recently, in the mid-twentieth century, did mechanical analogue computers come to be replaced by electronic analogue ones with resistors, capacitors, inductors and amplifiers. By the 1970s, digital computers had taken over for a variety of reasons, not least that mechanical machines were cumbersome. A more accurate answer meant bigger: because information is encoded in a physical way (think of the markings on a slide rule, for example) an extra bit of precision doubles the size of the device.[2] Compared to digital, mechanical analogue computers were also power hungry, slow and limited by the precision with which we can measure analogue quantities.

Unlike their purpose-built analogue predecessors, abstracting the input data from an analogue form into a string of 1s and 0s allowed digital computers to be reprogrammed to perform multiple types of calculations. Advances in integrated electronic circuits allowed digital computers to rapidly shrink their footprints and boost their power and speed. But analogue computing is making a comeback, not least because the approach can slash power consumption, which is increasingly an issue for the behemoths of computing. Analogue processing also offers a way to capture the chaotic richness of the real world.[3] Meanwhile, the rise of quantum computing offers extraordinary new opportunities for Virtual You.

Metamaterials

There are many ways to build analogue computers. One promising approach is through next-generation materials technology: metamaterials (*meta* from the Greek, meaning "beyond"). In 2006, Roger found himself in Imperial College London, examining a metamaterial that Sir John Pendry had developed with colleagues at Duke University, North Carolina, to cloak objects from radar. The confection of copper wires and loops on what looked like a pizza base realised an idea put forward by Pendry in 1999 (the name *metamaterial* was not coined until later).[4,5*]

Metamaterials present opportunities that are as extraordinary as their ability to manipulate light. One intriguing application is in cloaking. Metamaterials with a gradient of positive refractive indices can make light flow around an object, as in Pendry's "pizza." The same light-warping phenomenon can be seen in a desert mirage, when blue light from the sky is bent by gradients in heated air to create the impression of a pond in the desert.[6] While a metamaterial cloak hides an arbitrary object from view, rendering it invisible, oth-

---

* That work also stirred a huge amount of interest because it suggested metamaterials could be crafted to have a negative refractive index, unheard of in nature at that time but postulated decades earlier by the Russian physicist Victor Veselago after playing with Maxwell's equations.

ers are harnessing metamaterials to carry out analogue computing using light.

One such effort is under way in the Department of Electrical and Systems Engineering at the University of Pennsylvania, coincidentally the birthplace of ENIAC, the pioneering digital general-purpose computer. Nader Engheta, who is working on metamaterials with colleagues in Texas and Italy, believes that analogue optical processing is not only efficient but could be suitable to tackle problems that are intractable on digital machines.[7]

Roger first came across Engheta in 2005, when discussing cloaking with metamaterials, what he then called "transparency."[8] A decade later, Engheta showed how, instead of using bulky lenses and optics, a metamaterial one light wavelength thick could perform a suite of mathematical functions.[9] The input is encoded in the form of the complex shape of a light wave. The tailored properties of the metamaterial—notably the profile of refractive index, magnetic and electric properties—shape how this wave propagates, and Engheta could craft this wave to do something mathematically useful as it passed through the metamaterial, such as solve integral equations, at speeds orders of magnitude faster than their digital counterparts, while drawing less power.

His team demonstrated a proof-of-concept experiment using microwaves, which have a much longer wavelength than visible light (microwaves range from around 30 centimetres to 1 millimetre). The microwaves were manipulated with what he called the "Swiss cheese," a 2-square-foot block of polystyrene plastic with a distribution of air holes designed by Engheta to solve an integral equation with a given "kernel," the key part of the equation that describes the relationship between two variables. After the microwave passes through, the solution is represented by the shape, intensity and phase of the exiting wave. In 2021, he showed how another "Swiss cheese" could handle two different integral equations simultaneously, using two microwave frequencies.[10]

Because they manipulate microwaves, these devices measure about four wavelengths by eight wavelengths—that is, 30 cm × 60 cm in size, making them easy to troubleshoot.[11] At the time of writing, Engheta and his colleagues were working on a photonic chip a few

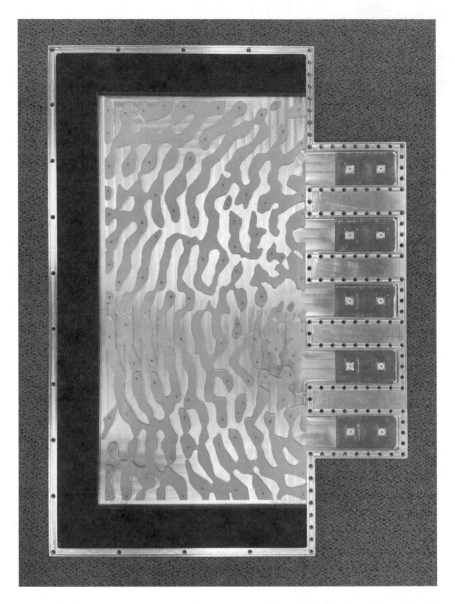

FIGURE 44. The pattern of hollow regions in the Swiss cheese is designed to solve an integral equation with a given "kernel," the part of the equation that describes the relationship between two variables. (Reproduced with permission by Nader Engheta)

microns across.* By using metamaterials that can alter their proper-ties, it might also be possible to program these devices, akin to the way laser light writes information on old-fashioned CDs. Parallelism is also possible by using different wavelengths—colours—simultaneously.

There are other possibilities for optical computers. As one ex-ample, Natalia Berloff at the University of Cambridge showed how to multiply the wave functions describing the light waves instead of adding them, offering new opportunities to solve computationally hard problems.[12] We also encountered physical neural networks in Chapter Four, where light passing through any physical object—such as a glass paperweight—could be used as the central processor in a fast and efficient deep neural network. Whatever the approach to optical computing, it seems likely that next-generation exascale machines will have analogue subsystems to do what Engheta calls "photonic calculus."

Silicon Brains

The human body itself can help spur the development of Virtual You, through the rise of analogue computers that have been inspired by its analogue workings. Building on pioneering work of the American scientist and engineer Carver Mead in the late 1980s, a field called neuromorphic computing has drawn on the brain as the inspira-tion for the design of computational hardware.[13] In one promising approach to neuromorphic engineering, Stan Williams from Texas A&M University, Suhas Kumar of Hewlett-Packard Labs and Ziwen Wang of Stanford University created a nanoscale device consisting of layers of different inorganic materials, notably of niobium dioxide,[14] that could recapitulate real neuron behaviour.†

Their work dates back to 2004, when Williams became fascinated by the work of Leon Chua at the University of California, Berkeley, an electrical engineer who was the first to conceive of the memristor. You

---

* Nader Engheta, email to Roger Highfield, March 18, 2021.
† Stan Williams and Suhas Kumar, interview with Peter Coveney and Roger Highfield, October 2, 2020.

can think of this electronic component as a "resistor with memory," as it remembers the amount of electric charge that flowed through, even after being turned off. As in your head, processing and memory lie in the same adaptive building blocks, so that energy is not wasted shuffling data around.

Postulated by Chua in 1971 as a "missing link" in circuits that traditionally used only capacitor, inductor and resistor,[15] then actually found by Williams and colleagues in 2008,[16] this fourth "atom" of all electrical circuits has been shown to be at least 1000 times more energy efficient than conventional transistor-based hardware used in artificial intelligence, for example.[17]

Chua draws parallels between the behaviour of memristors and neurons. In particular, he argues that sodium and potassium ion channels can be thought of as memristors, which are the key to generating the action potential seen in the Hodgkin-Huxley equations.[18] Chua reformulated them using four differential equations, or state variables, each of which is a dynamical system and nonlinear too, hence the ability to generate spikes, or action potentials. When he explored the behaviour of Hodgkin-Huxley, he found a transition zone between order and disorder, dubbed "the edge of chaos" (Chua describes it as "between dead and alive"*), which he believes is the key to the emergence of complex phenomena, including creativity and intelligence. (We discuss the edge of chaos in our book *Frontiers of Complexity*.)

Williams, inspired by Chua's papers, began to create devices to explore this nonlinear regime. First, he created an experimental memristor in 2006. Then he wanted a device that mimicked the Hodgkin-Huxley equations. By 2020, Williams, Kumar and Wang had come up with a synthetic neuron, a third-order memristor-based switch on a layer of niobium dioxide. When a small voltage is applied, the layer warms and, at a critical temperature, turns from an insulator to a conductor, releasing stored charge. This creates a pulse that closely resembles the action potentials produced by neurons, after which the temperature falls and the device returns to its

---

* Leon Chua, 7th Memristor and Memristive Symposium in Catania, Italy, October 1, 2021.

resting state and begins to reaccumulate charge. Just as Chua had shown that three differential equations were enough to capture the behaviour of an axon, so they could represent three distinct differential equations—or three state variables—in an inorganic structure: by changing the voltage across their synthetic neurons, they could create a rich range of neuronal behaviours. They created 15 in all (they believe there are 23, based on studies of a "fourth-order" system) that have been observed in the brain, such as sustained, burst and chaotic firing of spikes. These "electro-thermal memristors" can also explore the "edge of chaos."[19]

They went on to show that networks of synthetic brain cells can solve problems in a brain-like manner, using a network of 24 nanoscale devices to tackle a toy version of a real-world optimisation problem that would require a lot of energy for a conventional computer to carry out: the viral quasi-species reconstruction problem, where mutant variations of a virus are identified without a reference genome.

The devices were wired together in a network inspired by the connections between the brain's cortex and thalamus, involved in pattern recognition. Depending on the data inputs, the researchers introduced the network to short gene fragments. By programming the strength of connections between the artificial neurons within the network, they laid down rules about joining these fragments. Within microseconds, their network of artificial neurons settled down in a state that was indicative of the genome for a mutant strain. Their analogue brain–inspired devices were at least 10,000 times more energy efficient than a digital processor for solving the same problem.

In Europe, neuromorphic computing is part of the billion-euro Human Brain Project that involves more than 500 scientists and engineers at more than 140 institutions. SpiNNaker (spiking neural network architecture) is one of the largest neuromorphic supercomputers and has been under development since 2005 by a team led by Steve Furber at the University of Manchester, England.[20] This machine connects a million processors that have been simplified to cut power consumption and heat generation (these processors, made by the company ARM, are the kind that have enabled phones to be mobile) with a network optimised for the transmission of action

FIGURE 45. The SpiNNaker brain-like computer. (Reproduced with permission by the University of Manchester)

potentials—as a series of spikes—to processors in a ring-doughnut-shaped network. They have modelled various brain regions, which could in theory be interlinked to create a larger brain model.* A new version is under development, SpiNNaker2, although these machines are not analogue. "All neuromorphic systems—even the ones that use analogue neuron and synapse processing—generate digital outputs in the form of spikes," Furber told us.†

The Human Brain Project includes another neuromorphic machine, BrainScaleS, at the University of Heidelberg, where the effort is led by Johannes Schemmel. The first generation of BrainScaleS consisted of four million artificial neurons and one billion synapses on 20 silicon wafers, again communicating using digital spikes to efficiently mimic the biological equivalent, notably the way a brain

---

* Steve Furber, interview with Roger Highfield and Peter Coveney, October 30, 2021.
† Steve Furber, email to Roger Highfield and Peter Coveney, November 18, 2021.

FIGURE 46. The BrainScaleS-1 machine in Heidelberg, Germany. (Reproduced with permission by Heidelberg University)

learns by changes in local connections rather than a global algorithm. The team created BrainScaleS to show that they could connect analogue circuits on a large scale to study brain information processing, where information is shared by digital spikes running in analogue time, backed up by data from independent studies of real brains (of mice, rats, cats and monkeys), along with mathematical modelling, high-performance computing and analysis.

At the time of writing, the team was working on BrainScaleS-2, which in its final expansion should use up to 1000 wafers that are double the size of those in its predecessor and are based on deeper understanding of flexible learning rules and neural architecture. The result will be a true analogue neuromorphic computer that is capable of emulating "small brains."* However, although they promise to be

* Johannes Schemmel, interview with Peter Coveney and Roger Highfield, November 23, 2021.

quicker and use significantly less energy than a digital computer, Schemmel does not expect these analogue devices to offer greater precision than digital ones in brain emulation.

There are many more examples. One effort led by Giacomo Indiveri at ETH Zurich has created a neuromorphic chip along the lines first envisaged by Carver Mead that reliably recognizes complex biosignals. There is the TrueNorth digital chip, developed at the IBM research laboratory in Almaden, California; Intel's Loihi digital platform, designed to encourage neuromorphic research, Stanford University's Neurogrid board; BrainChip's Akida chip; GrAI Matter Labs' GrAI One chip for machine learning; and a 1024-cell electronic synapse array, based on memristors, developed to classify faces in a more efficient way by the Institute of Microelectronics, Tsinghua University and Stanford University.[21]

To accommodate both digital computer-science-based machine learning algorithms and analogue neuroscience-oriented schemes, such as brain-inspired circuits, the Tianjic chip was devised by researchers from Tsinghua University, Beijing Lynxi Technology, Beijing Normal University, Singapore University of Technology and Design and the University of California, Santa Barbara.[22] Using just one chip, containing approximately 40,000 "neurons" and 10 million "synapses," Luping Shi of Tsinghua and colleagues demonstrated the simultaneous processing of algorithms and models in an unmanned bicycle, delivering real-time object detection, tracking, voice control, obstacle avoidance and balance control. Using the chip, an autonomous bicycle can not only balance itself but steer around obstacles, respond to voice commands and make independent decisions. A hybrid chip inspired to some extent by brain-like principles can indeed ride a bicycle.

Quantum You

Another kind of calculating machine now beckons: the quantum computer. These computers are predicted to have remarkable properties that, if realised, could also accelerate the development of Virtual You.

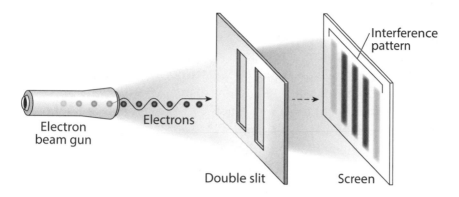

FIGURE 47. The double slit experiment, using electrons. (Adapted from Johannes Kalliauer; Wikimedia Commons, CC0 1.0)

The reason is down to how, in the quantum domain of atoms and molecules, reality is described by scientists in terms of mathematical objects called wave functions, which present profound new opportunities for computation, and in turn for Virtual You. These functions, in effect, contain all possibilities of what we could encounter when we make an observation. However, once you make a measurement you get only one result. In one textbook example, this could be how particles of light—photons—create an interference pattern of bright and dark stripes on sensors after they have passed through a double slit. The pattern indicates that each particle is actually a wave that passes through both slits at once, producing two wavefronts that converge and interfere, brightly reinforcing one another in some places and darkly cancelling in between. Each particle materialises with highest probability at the regions of most brightness, but, remarkably, if you add a second sensor to detect which slit each particle passes through, the interference pattern disappears. The same goes for electrons.

In 1926, the German physicist Max Born (1882–1970) came up with an interpretation of this result that indicates that the root of all reality should be couched in terms of probabilities rather than certainties. The following year, his fellow countryman Werner Heisenberg (1901–1976) published his uncertainty principle, which states that accurately measuring one property of an atom puts a limit to the precision of a measurement of another property.

Born's interpretation is more radical than it sounds and provides insight into why quantum mechanics is so peculiar: the variables in Newton's equations refer to objective properties, such as the mass or velocity of a cannonball, but, according to Born's postulate (which, by the way, we are still working out how to justify[23]), the wave function of a particle of light, a photon, shows what outcome we might get were we to make a measurement.

At the moment of observation or measurement, when a photon is detected, the wave function collapses to actuality, according to the "Copenhagen interpretation," developed by the Danish scientist Niels Bohr (1885–1962). This seems to contrast with the traditional interpretation of "classical" physics as being strictly deterministic (though, as we have seen, dynamical chaos blurs the notion of what we mean by determinism).

Despite the importance of his wave equation, Schrödinger is more widely known in popular culture for a thought experiment he devised in 1935 to show why the description of the atomic world given by quantum mechanics is downright odd; it also helps to show why today's quantum computers have such unusual properties.

Schrödinger imagined a cat trapped in a box, along with a radioactive atom that is connected to a vial containing a poison. If the atom decays, it causes the vial to smash and the cat to perish. When the box is closed, however, we do not know if the atom has decayed or not, which means that it exists in a combination of the decayed state and the nondecayed state. Until someone peeks inside, the cat is neither alive nor dead; it is in what is called a "superposition," both alive and dead. Only at the moment we crack open the box and take a peek—that is, conduct a measurement—does the wave function "collapse" into one actuality. Similarly, in a quantum computer, bits consist not of 0s or 1s, but of qubits that are 0s *and* 1s—until the answer is sought.

There are other ways to interpret what we mean by a "measurement." One radically different view was published in 1957 by the American physicist Hugh Everett III (1930–1982): the universe is constantly splitting into myriad new ones, so that no collapse takes place. Every possible outcome of an experimental measurement occurs, with each one being realised in a separate parallel universe. If one accepts Everett's interpretation, our universe is embedded in an ever-multiplying mass of parallel universes. Every time there is an event at the quantum level—a radioactive atom decaying, for example, or a particle of light impinging on your retina—the universe is supposed to "split" into different progeny. The English physicist Paul Davies once quipped that Everett's startling idea is cheap on assumptions but expensive on universes.[24]

Notwithstanding these debates about how to make sense of quantum mechanics, the theory has proved successful at describing the atomic domain, and in the 1980s it began to dawn on scientists that the theory also had implications for computing: Alan Turing was working with classical physics and, as a consequence, his universal computer was not so universal after all. By extending Turing's groundbreaking work, Richard Feynman,[25] Paul Benioff and Yuri Manin laid the groundwork for quantum computing.

In 1985, the British physicist David Deutsch proposed a universal computer based on quantum physics, which would have calculating powers that Turing's classical computer (even in theory) could not simulate.[26] Almost a decade later, the mathematician Peter Shor showed while working at Bell Labs that calculations that would take a classical (that is, nonquantum) computer longer than the history of the universe could be done in a reasonable time by a quantum computer.[27] Around that time at the Los Alamos National Laboratory, Seth Lloyd, now a professor at MIT, came up with the first technologically feasible design for a working quantum computer.[28]

Quantum computers offer extraordinary possibilities for Virtual You because they rely on a fundamentally new way to process information. Information takes the form of bits, with a value of 1 or 0, in the "classical" computers that inhabit our homes, offices and pockets. Instead, quantum computers manipulate qubits, which can represent both a 1 and a 0 at the same time. Like Schrödinger's cat, the 1 and 0 are in a "superposition" that is neither 0 nor 1.

Quantum computers should in principle be ideal for chemistry and biochemistry, as Feynman had suspected, because quantum mechanics is used to work out the way electrons dance around a molecule (to calculate its electronic state). To do this, one has to deal with what is called Hilbert space, the abstract arena where quantum information processing takes place. This is difficult on a classical computer because the Hilbert space expands exponentially as one increases the size of the molecule, in particular the number of electrons and orbitals it contains. This poses particular challenges for simulating the chemistry of the cell, where the workhorses are all very large molecules—proteins.

The algorithmic complexity of the most accurate electronic structure calculations scales factorially with the number of orbitals (this is worse than exponentially, growing by multiplying with an increasing amount, not a constant amount). By comparison with a classical computer, a relatively small number of qubits in a quantum computer can accommodate this enormous storage requirement. Indeed, the electronic structure calculations scale only linearly with the number of qubits and the wave function is distributed over them all; whereas, on a classical computer, you would need an exponentially growing memory, which is beyond the reach of any such machine.

As a consequence of superposition, two distant parts of a large system that one might expect to be unconnected are in fact correlated because of the nonlocal nature of quantum mechanical interactions. Such interlinked behaviour is called "entanglement." This was one reason why Albert Einstein, one of the architects of quantum theory, was uncomfortable with its implications. He realised that if you separated the entangled part of such a system, even by a vast distance, it would still be described by the same wave function and thus would remain instantaneously correlated. Today, we know that this "spooky action at a distance," as he called it, does indeed occur in quantum systems.[29]

Quantum computing is based on the premise that we can maintain these entangled superpositions of states during quantum parallel computing and get the answer when a measurement takes place. In this way, a quantum computer orchestrates superposition, entanglement and interference to explore a vast number of possibilities simultaneously and ultimately converge on a solution, or solutions, to a problem.

New opportunities beckon. In the journal *Science*, Peter Love, now at Tufts University, Alán Aspuru-Guzik, now at the University of Toronto, and colleagues showed the potential of quantum computing to solve large electronic structure problems, where the electronic structure of a molecule consists of the collective behaviour of the electrons that bind the molecule together.[30] Solutions will enable chemists to calculate the energy required to make—or break—a molecule, which is crucial when working out how well a drug might work in the body, for example.

However, a molecule's electronic structure is hard to calculate, requiring the determination of all the potential states the molecule's electrons could be in, plus each state's probability. Since electrons interact and become entangled with one another, they cannot be treated individually. With more electrons, more entanglements arise, and the problem gets at least exponentially harder. Exact solutions do not exist for molecules more complex than the two electrons found in a pair of hydrogen atoms.

If quantum computers are able to work out the electronic structures of more complex molecules than molecular hydrogen, they could come up with an "exact" description of the interaction of candidate drug molecules with target proteins (though, of course, it might be exact in terms of the molecules and targets but not in terms of conditions found in the crowded cells of a patient). As we saw with multiscale modelling, one can blend this approach with classical modelling outside the quantum part to make best use of limited qubits. Because many groups, such as Peter's at University College London, are also exploiting machine learning to accelerate the drug discovery cycle, there is some excitement that quantum machine learning could turbocharge this endeavour.

In a similar way, quantum computers could accelerate the development of Virtual You by modelling the enzymatic reactions that take place in the human body. One would again seek to use quantum computing to speed up electronic structure reactions at the heart of the enzyme's active site, leaving the rest of the simulation to classical computing. That would allow faster modelling and simulation of how enzymes function and thus better understanding of the large-scale chemical reaction networks that dominate the chemistry inside our cells.

Rise of the Quantum Machine

The power of quantum machines continues to grow, though their potential to contribute to Virtual You is still some way off. In 1998, Isaac Chuang of the Los Alamos National Laboratory, Neil Gershenfeld of the Massachusetts Institute of Technology and Mark Kubinec of the University of California, Berkeley, created the first (2-qubit) quantum computer that could be loaded with data and would then output a solution. Two decades later, IBM unveiled the first 50-qubit quantum computer, the appearance of which was likened by some commentators to "a steampunk chandelier." By 2021, one Chinese manufacturer, Shenzhen SpinQ Technology, had announced the first desktop quantum computer, albeit with only 2 qubits, and IBM's Eagle processor hit 127 qubits. However, to realise Feynman's original ambition to do serious quantum chemistry will require quantum computers that possess *hundreds of reliable qubits.*

Quantum computers come in many guises because any two-level quantum-mechanical system can be used as a qubit, from the spin of an electron or nucleus to the polarisation of a photon. There are many different kinds of quantum computers, depending on the types of qubits, how you manipulate the qubits, and how you entangle atoms. As Seth Lloyd likes to put it, it is all down to how you want to empower atoms.*

However, there are many technical hurdles to overcome in quantum computing. The most fundamental is the fragility of qubits, which must be kept in a coherent state of superposition or else they will collapse due to measurement-like processes. Any environmental interference in the quantum system—"noise," caused by variations in temperature, vibration, electromagnetic waves—produces what is known as decoherence, where qubits lose their useful ambiguity and crystallise into humdrum 1s and 0s. This stability is measured in "coherence time," which reflects how long a qubit survives before it collapses into either a 1 or 0.

The longer the coherence time, the longer a quantum computer has to complete its calculations. At the moment, this is fleeting, at

---

* Seth Lloyd, interview with Roger Highfield, May 22, 2018.

around 100 microseconds; the slightest disturbance, such as a vibration or change in temperature, results in persistent and high error rates when executing an algorithm. As a result, today's quantum computing devices and those of the foreseeable future are referred to as Noisy Intermediate Scale Quantum (NISQ) computers and are highly unreliable, meaning that one has to repeat measurements and accept less-than-perfect results. The more qubits that are added to these devices to solve harder problems, and the more complicated and deeper the quantum circuits become, the more difficult it is to maintain the tenuous state of coherence, level of error mitigation and noise reduction necessary for them to yield meaningful answers.

However, Peter and his PhD students Alexis Ralli, Michael Williams de la Bastida and Tim Weaving have found a way to focus the power of today's small NISQ computers, as well as those that will come beyond them, on the tougher part of the problem, hiving off the easier regions of the calculation to a classical computer. This pragmatic step forward, akin to the adaptive mesh refinement we encountered in Chapter Seven, improves the accuracy of results for molecules that currently lie outside the reach of full quantum computation, which is the case for essentially any of chemical and biological interest, from drugs to proteins and DNA.[31]

There is, however, a more fundamental issue with using quantum computing to simulate Virtual You: we still do not understand how wave functions collapse, which is what leads to the decoherence referred to above. As Roger Penrose remarked, this process is not explained in current quantum mechanics and, in his view, is nonalgorithmic. One could resort to the alternative interpretation preferred by David Deutsch, for example, though disliked by many more—that the universe splits into multiple realities with every measurement—but the deeper point is that there is no generally accepted theory of how a hazy probabilistic quantum entity morphs into a classical certainty.

What actually corresponds to a "measurement" in quantum theory is still not clear. It is troubling that so much effort is being expended to maintain quantum coherence in quantum computers without any agreement on knowing what causes it to be lost. This is the most severe of the major constraints that limit the realisation of quantum computing's extraordinary potential. Pragmatists, such as Peter Love, claim that this should not matter, just as it does not matter to the legions of people who rely on classical computing that we don't know why electrons have a given charge.* But this issue is fundamental: quantum theory remains incomplete without a scientific theory of wavefunction collapse and, as a consequence, there is a missing element at the heart of quantum computing.

Quantum Advantage

Despite these caveats, many regarded it as a milestone in 2019 when Google scientists announced that they achieved the feat of "quantum supremacy,"[32] a term coined some years ago that today is disliked by many, for obvious reasons, who prefer the less triumphalist term "quantum advantage."

For their "beyond classical experiment," the Google team used a quantum computer called Sycamore, which relies on superconducting circuits, that is, circuits that conduct electricity without resistance at low temperatures. Even though they had relatively modest numbers of qubits—53—they could explore a vast computational state space, the Hilbert space we encountered previously. The "advantage" came from the realisation that their Sycamore processor could tackle what the Summit supercomputer at Oak Ridge National Laboratory in Tennessee would take millennia to do. The simulations took 200 seconds on the quantum computer when the same simulations on Summit would, they estimated, have taken more than 10,000 years to complete. Not only was Sycamore faster than its classical counterpart, but it was also approximately 10 million times more energy efficient.[33] Some claimed the feat was comparable in importance to the Wright brothers' first flights.[34]

The crux of the demonstration was a problem that was hard for a classical computer but easy for a quantum computer: the random circuit sampling problem. Related to random-number generation, the task involved sampling the output of a quantum circuit. Because

---

* Peter Love, email to Peter Coveney and Roger Highfield, November 23, 2021.

FIGURE 48. The Sycamore processor mounted in a cryostat to keep it cool. (© Google)

of quantum entanglement and quantum interference, repeating the experiment and sampling a sufficiently large number of these solutions results in a distribution of likely outcomes.

But this task is extremely artificial. Yes, a quantum computer seemed to have won a race with a classical supercomputer and, yes, the fidelity of its calculations was impressive. But the practical implications are minimal, and it remained possible that classical algorithms could catch up because this quantum advantage was over current optimal classical algorithm strategies. Soon after the announcement, IBM proposed a type of calculation that might allow a classical supercomputer to perform the task Google's quantum computer completed. Then, in 2022, a Chinese team led by Pan Zhang of the CAS Key Laboratory for Theoretical Physics in Beijing claimed an exaflop machine running its algorithm would be faster than Google's quantum hardware.[35]

Like the Wright brothers' Wright Flyer, which once sat in the Science Museum (until a feud between Orville Wright and the Smithsonian Institute had ended, when it returned to the United States in 1948), this milestone was not really offering a practical solution to any problem but rather suggested that a new technology was close to lift-off.

A more convincing example of quantum advantage was reported by a team in China in 2020, using a room temperature quantum computer named after an early mathematical work in China's long history entitled *Jiuzhang suanshu* (Nine Chapters on the Mathematical Art), a second-century BCE text compiled by generations of scholars, where *Jiu* represents "9" and, by dint of being the largest digit, also means "supreme."[36]

Developed by a team led by Chao-Yang Lu and Jian-Wei Pan at the University of Science and Technology of China, Hefei, the quantum computer cracked a problem that seemed beyond classical computers. Japan's Fugaku supercomputer, at that time the world's most powerful classical computer, would take around 600 million years to accomplish what Jiuzhang could do in just 200 seconds.*

---

* Chao-Yang Lu, email to Roger Highfield, December 28, 2020.

FIGURE 49. The quantum computer Jiuzhang 1.0 manipulates light using an arrangement of optical devices. (© The University of Science and Technology of China)

Unlike Sycamore, Jiuzhang is an analogue machine, one designed for boson sampling, where bosons are one of the basic classes of particles, which includes photons. Boson sampling entails calculating the probability distribution of many photons after they encounter a device called a beam splitter, which divides a single beam into divergent beams. This is the quantum measurement problem, and there is no mathematical equation that accounts for this collapse, which Roger Penrose claims to be noncomputable.

Jiuzhang sent laser pulses into a maze of 300 beam splitters, 75 mirrors and 100 detectors capable of spotting a single photon (the fact that photons are being counted does lend a digital aspect to the way Jiuzhang works). With enough photons and beam splitters, Jiuzhang could produce reliable interference patterns, a distribution of numbers of photons that—because the problem scales exponentially—lies beyond the reach of a classical computer. Because Jiuzhang used photons as bosons, not as qubits, its power cannot be translated into qubits for comparison with digital quantum computers.

One advantage of digital computers—and Sycamore too—is that they can be flexibly programmed to do different things while Jiuzhang is hardwired only to do boson sampling. However, Jiuzhang is being upgraded,[37] and the team believes that boson sampling can also be adapted for machine learning,[38] quantum chemistry[39] and solving other problems to do with networks (using what mathematicians call graph theory[40]).

Compared with conventional computers, these photon-based devices offer other advantages. They have higher throughput and work at the speed of light. Like the computer between our ears, they also consume a fraction of the power of a classical computer. The power of these machines is growing fast. At the time of writing, the California company PsiQuantum was working on a photonic quantum computer with fault tolerance and error correction, along with at least 1 million qubits.

Quantum Exascale Analogue You

We do not expect quantum computers to sweep away classical machines. Conventional exascale, analogue and quantum machines have different attributes, strengths and limitations, all of which will make a contribution to Virtual You. The next generation of supercomputer is likely to be the best of all worlds, so, for example, an analogue "module" would be used to simulate nonlinear, chaotic phenomena, while a quantum "module" would be brought to bear on the molecular parts of a simulation.

One can already see modest hybrid computing solutions, such as an algorithm called the variational quantum eigensolver (VQE), which offloads the bulk of the computation of the electronic structure of a molecule to existing classical computation, while using a quantum computer for the hardest part of the calculation—changing bonds and other "moving parts" of chemical processes that are beyond the capacity of the biggest and fastest supercomputers.[41,42] However, it is important not to get carried away: VQE is intrinsically limited to tackling simple molecules of limited interest that are typically made of two or three atoms, such as lithium or beryllium hydride, and water.[43]

In other words, VQE is very much a provisional fix in the era of small and unreliable quantum devices. Even so, given the achievement of Jiuzhang, the possibilities of hybrid quantum-classical machines in the exascale era are intoxicating. In the case of Virtual You, quantum processors would be ideal for simulating the cosmic number of molecular interactions within your cells, while classical computers would best be deployed to simulate the larger-scale workings of the body, and analogue processors could be brought to bear on the brain, along with physiology where chaos and nonlinearity lurk. Importantly, they would be much more energy efficient than a digital processor.

When a virtual version of you does eventually take its first breath, it will rely on both analogue subsystems that operate at nearly the speed of light, and draw little power, and quantum modules that harness ideas that have perplexed us for more than a century. Will we, in a matter of moments, be able to use them to test a range of drugs on a virtual copy of ourselves, or use them to work out the healthiest way to live? Will these machines one day be able to forecast our future faster than reality unfolds?

From Healthcasts to
Posthuman Futures

*He will manage the cure best who has foreseen what is to happen from
the present state of matters.*

—Hippocrates*

In one important sense, the urge to create Virtual You is as old as
humanity. Drawings, carvings and other representations of human
form date back long before Rembrandt's endless depictions of his
favourite model (himself), before medieval anatomists became fascinated by what lies under our skin, and millennia before the rise of
anatomy in ancient Egypt.

Our preoccupation with depicting our bodies is so old that it is
almost instinctive, providing a distorted mirror through which to
view the agenda of ancient minds. In important ways, their concerns
were not so different from our preoccupations today, which also
revolve around life, health and death.

The earliest known drawing by *Homo sapiens* dates back more than
70,000 years and features nine strokes of red ochre—six parallel
lines diagonally crossed by three curved lines—on a stone flake.[1]
Perhaps self-representation emerged with the birth of the modern
mind during the Stone Age, at least 300,000 years before the present
day.[2] Perhaps, for that to occur, we had to see ourselves as somehow
separate from the natural world, according to the British author
Jo Marchant. She points out how in Lascaux, a complex of caves
in France's Dordogne, the art of 17,000 years ago focused mostly

---

* From "The Book of Prognostics," part 1 (400 BC), in *The Genuine Works of Hippocrates*
(trans. Francis Adams), vol. 1, 113 (1849).

on aurochs, horses and deer, with a few stencils of hands and stick figures. Later, in the Neolithic era and with the rise of agriculture, there is a proliferation of human representations, which suggests rising self-awareness and a growing chasm between us and nature.[3]*

These primitive twins spread worldwide. When modern humans turned up on the Indonesian island of Sulawesi 50,000 years ago, they left traces behind in the form of rock paintings, reddish-orange hand stencils and carved figurines.[4] Another outpouring of figurative art came during the migration of *Homo sapiens* into central and western Europe, where they would eventually displace the resident Neanderthals, around 30,000 years ago. One famous example, found in the Hohle Fels Cave in southwestern Germany, was a squat and full-figured—albeit headless (some believe it shows the parts of the body that could be seen)—torso measuring less than 2.5 inches long with projecting breasts and explicit genitals, carved in a mammoth tusk at least 35,000 years ago.[5]

Art seems so old and so ubiquitous that ancient peoples must have drawn some comfort from depicting the human form. But what? Perhaps it was the sheer delight of seeing these primitive representations as they were animated under the guttering light of a fire or tallow candle. Perhaps they were the Stone Age equivalent of a selfie.[6] Or perhaps they were linked with health, notably fertility and the most urgent biological imperative of all, to reproduce.

No doubt these rituals emerged in the face of an uncertain world as our ancestors had to wrestle with all manner of threats, some real and others figments of the imagination. Roger has written on how sun worship—along with proto-Christmas celebrations—was born in the Northern Hemisphere when, during bitter winter months, our ancestors worried that the sun might ebb away, even disappear.[7]

Ultimately, the concerns that drive the creation of digital twins and the creation of Virtual You are the same as those that furrowed brows in primeval times: survival. Once upon a time a terracotta model of a damaged or diseased body part was used as a votive gift to summon divine attention for a pressing medical problem. Just as our ancestors wanted to capture something of their identity, so the

---

* Jo Marchant, interview with Roger Highfield, September 2, 2020.

quest for Virtual You is also driven by individualism, curiosity about the future and a fundamental urge to live.

The impact of digital twins might be felt far beyond medicine. Aside from permitting better-informed lifestyle choices, they offer all sorts of other possibilities, from policymaking to virtual reality, gaming and personal entertainment. When it comes to the latter, AI and human simulations that are two-dimensional are already being used by Hollywood, for customer service, public outreach and more. Digital people will undoubtedly become more convincing and influential when, thanks to Virtual You, their programming is more than skin deep. While current video games use a physics engine to allow their characters to experience gravity, inertia, collisions and so on, future games will have a "medicine engine" to simulate how much harm comes to players during violent encounters.

Virtual You is part of a revolution in how we represent the world. While once we drew with ochre, carved mammoth ivory and projected flickering shapes on the walls of caves, there is now an extraordinary effort to recapitulate the features of the natural world *in silico*. Scientists and engineers are creating great computing behemoths that can manipulate oceans of electrons in microchips and make them dance upon a screen, where myriad forms can appear in

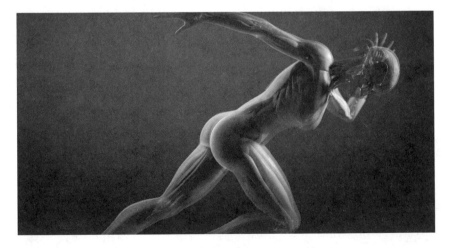

FIGURE 50. Still from the *Virtual Humans* movie. (CompBioMed and Barcelona Supercomputing Centre)

the guise of quantum and phosphor dots, liquid crystals and light-emitting diodes. Today, they are playing with ways to write information onto the retina, or into the brain.

Thanks to the five steps outlined earlier, we can use computers to simulate our world across a vast range of domains, from inside the nucleus of an atom to the large-scale structures that stretch across the universe itself. With the rise of data visualisation, we can show great expanses of dark matter, reveal a black hole's dark heart ringed in light that's distorted like a fun-house mirror, or demonstrate the workings of the molecular machines that run our cells.

We can model the extraordinarily complex domain that lies between atoms and galaxies, the mesoscale that human beings inhabit. With augmented and virtual technologies, we can even inhabit these virtual worlds and undertake fantastic voyages into our own organs, tissues and cells. The era of Virtual You is upon us, and it is vital that we begin to consider the moral and ethical implications of this rapidly approaching future.

Rise of the Digital Twins

Digital twin technology is now becoming established in medicine, and, as we have shown in the previous chapters, there are already glimpses of its potential, from trialling drugs with virtual heart cells to replacing animals in research, planning delicate brain surgery, honing the design of implants, speeding up the testing of implants and medical devices and ensuring that particles of an inhaled drug reach the right place in the body.[8]

We can already use a supercomputer to examine the enzymes and other molecular machinery in your cells to customise a drug so that it works best for you. Your virtual avatar will not only be able to react to drugs in the way that you do, it will be able to move like you too. We will be able to use it to calculate the forces on your bones, and predict the risk you will suffer a fracture if you have metastases or osteoporosis. More sophisticated insights will emerge as virtual organs are coupled together by virtual nervous systems, hormones and circulatory systems.

In the short term, Virtual You will be digital and, for the reasons explained in the previous chapter, will not quite mimic everything that you can do. Rather than be a virtual person, digital twins will represent a patient's many subsystems, from enzyme to cell to organ to limb. In many cases, we remain largely in the era of low-dimensional models built on simple theoretical foundations, that is, ordinary differential equations. In other words, our virtual creations often inhabit a bland flatland of mathematical possibilities, though this is still hugely valuable because it will help put medicine on a firmer scientific base.

We are entering the era of colossal data, thanks to new medical imaging methods, not to mention genomics. With the judicious use of machine learning, guided by theory, we can make more sense of these data. When married with physics-based understanding, and expressed with partial differential equations that are able to capture details of form and function, we can produce high-fidelity models of ourselves. With the exascale computing era, and new kinds of computing, notably quantum and analogue, we will be able to simulate more processes within the body, and in more baroque detail than ever possible before.

The path to Virtual You will be long, rocky and often incremental. Because of the limitations of digital twins outlined in Chapter Two, it will never culminate in a sentient being in cyberspace. Even so, as more features of people are successfully simulated, along with key elements of their environment, medicine will enter the four-dimensional, high-fidelity, full-colour era of Virtual You and exert a growing influence on real life. When all five steps outlined in this book are taken, the impact on medicine will be profound.

Current medicine is akin to a glorified lookup table, where you are diagnosed and treated depending on the past experience of doctors who have treated similar patients over generations. In future, medicine will increasingly be led by scientific insights into health and treatments that, akin to engineering, are based on theory, data, modelling and insights about how your own body works. Rather than always looking backward at the results of earlier clinical trials, medicine will become truly predictive.

Twenty-First-Century Medicine

There is much talk today of personalised medicine. However, no matter how much doctors say they want to treat us all as individuals, the current generation of medical treatments remains, roughly speaking, one size fits all. As we have emphasised throughout, current medicine is tantamount to staring in the wing mirror rather than gazing ahead out of the windscreen. Much faith is placed in AI and big data methods, but, though valuable, often they are only as good as the data they are trained on, and those data are frequently all about what happened to other people in the past.

In important respects, medicine has lost momentum. On average, each new one-size-fits-all drug costs around a billion dollars to develop,[9] takes over 10 years to bring to market and—at best—works for half of the population, roughly speaking (see, for example, studies of antidepressants,[10] neuraminidase inhibitors[11] and migraines[12]). The next time you get a prescription from your doctor, remember that most common medications have small to medium effects.[13] Sometimes you will be successfully treated by a given drug, often you will be left untouched. A few times you will develop harmful side effects.

The same goes for many other treatments, vaccines and so on. Peanuts are part of a balanced diet for some, yet deadly to others with nut allergy. An operation that goes smoothly for one person ends in disaster, even death, for another. Because we're all so different, the current generation of medical treatments is bound to fail. In decades to come, this generic approach will be seen as primitive.

The way we diagnose disease is mostly sporadic and often too late, as it largely relies on detecting symptoms. Hood and Venter have shown one way forward, by subjecting people to batteries of tests. We can speed up diagnosis by giving a regular update on the physiological state of a person, which is now becoming possible thanks to wearable devices and smartphones. Machine learning can also be trained to look for warning signs of health problems. But what we really need is advanced warning of possible problems. We need to take all these data and, using a predictive model, suggest actions to avert the development of symptoms in the first place.

In this sense, the current practice of medicine is surprisingly unscientific. There is a dearth of theory. Peter Hunter, who has spent decades working on virtual organs, was prompted to remark: "It is sobering that anatomy, physiology and biomedical science, which underlie the practice of medicine, are the only scientific disciplines that have not had a theoretical framework within which to assemble the vast knowledge of human biology that has been accumulated over hundreds of years. Therefore, these disciplines lack the means to rationally and systematically design experiments to fill the gaps in our knowledge." As we and others have argued, the time is right for a more systematic and multidisciplinary effort to develop biological theory.

Despite these limitations, progress has been made in predicting future health using various kinds of -omics, notably from the human genome programme. Decades ago, there was an expectation that one gene has one physiological function. The reality, of course, is that the physiological whole is much greater than its genetic parts.* It is closer to the truth to say that any bodily function depends in one way or another on all 20,000 genes as a result of all the cross talk in the body between many chemical pathways, each of which involves many genes. Some even talk about "omnigenic" disease, where "gene regulatory networks are sufficiently interconnected such that all genes expressed in disease-relevant cells are liable to affect the functions of core disease-related genes."[14] As we saw in Chapter One, we have barely scratched the surface in terms of understanding DNA in scientific terms.[15] Yes, the genome is important, but, as a result of additional layers of biological complexity, from noncoding regions of DNA to epigenetics,[16] the impact of genomics and genomics-based "precision medicine," though significant, has fallen short of the hyperbole.

The first-draft human genomes were composites of DNA from several people, based on one set of chromosomes, not two, and neglected around 15% of the genome, which was beyond the sequencing methods of the day. When these genomes were unveiled during a White House ceremony in June 2000 by Craig Venter and Francis

---

* Peter Hunter, interview with Roger Highfield, October 23, 2020.

Collins, the then US president, Bill Clinton, talked about how genome science "will have a real impact on all our lives." A decade later, in 2010, one study concluded that "genetic screening of whole populations is unlikely to transform preventive health in the ways predicted 10 years ago."[17] A *Scientific American* article that same year was headlined: "Revolution Postponed: Why the Human Genome Project Has Been Disappointing." In 2020, Eran Segal of Israel's Weizmann Institute of Science pointed out that "genomics has yet to deliver on its promise to have an impact on our everyday life."[18]

Understanding the human genome was supposed to be a boon for drug development. But the vast expense and sluggishness of Big Pharma's increasingly unprofitable research and development pathway to new drugs shows that there is a long way to go to shape medicine in our own image. The need for novel antibiotics to deal with the relentless rise of resistant bacteria, or superbugs, is as pressing today as ever. When it came to the COVID-19 pandemic, it was striking how several vaccines were developed in less than a year, while the first wave of COVID-19 treatments were not antivirals but depended on pressing old drugs into service.[19,20]

Various people, such as Craig Venter and Leroy Hood, have been at the forefront of scouring personal data for early signs of disease, as mentioned earlier.[21] When it comes to traditional approaches to understanding human biology, Hood refers to the parable of the blind men and an elephant, who struggle to envisage what the elephant is like by touch alone.* In the past, doctors tended to reduce the colossal complexity of the body to exterior signs. Nor did they integrate all this information, or think much about how to assay internal complexity. The latter can be revealed by the results of protein and DNA sequencing, along with an assortment of other tests, for instance, that measure the rich array of chemical signals and metabolites secreted into blood. These insights into the health status of the body can reveal biomarkers of transitions from health to illness, along with the nature of disease. For example, when it comes to Alzheimer's, the devastating cause of dementia, Hood's data suggest there are six subtypes.

---

* Leroy Hood, interview with Peter Coveney and Roger Highfield, August 14, 2021.

Hood has developed ways to gather big data, integrate them into a "phenome" and predict "biological age" in combination with mechanistic models—effectively, an evaluation of how well you are aging.[22] This assessment suggests that Hood himself is 15 years younger than his chronological age, which he credits to diet and exercise. By comparison, some sufferers of COVID were aged by 20 years as a result of infection with the virus. He believes gathering and acting on big data—"longitudinal phenome analysis"—is the key to reducing biological age.

Using modelling, his colleagues have gleaned insights into the development of Alzheimer's in those with the most prevalent genetic risk—they carry a variant of the APOE gene, known as APOE4, which affects the ability of brain cells to metabolise lipids and respond to stress. Hood's team has also studied how to arrest the development of Alzheimer's, with early work suggesting exercise is much more effective than drugs (an insight welcomed by Roger, who has used a treadmill desk for more than a decade). In all, Hood believes well-being rests on a mixture of about 60% environmental and lifestyle factors, with genomics contributing about 30% and healthcare an unimpressive 10%.*

These and the many other elements of Virtual You discussed in earlier chapters are already having an impact on medicine. When digital twins are commonplace, they will offer richer opportunities for treatments than "precision medicine," which predicts how you will respond to a treatment based on the responses of others who are "like you," or machine learning approaches that have learned what is likely to happen to you based on how you have reacted in the past, or the responses of others who are similar to you.

Today, a diagnosis is sometimes barely more than a list of symptoms united by little understanding and an impressive name. Traditional patient records are fragmentary collections of data and symptoms, gathered during intermittent crises. Theory-based modelling offers a way to make sense of symptoms. Simulations can put pieces of patient data together over a vast range of length and timescales to work out how the bodily whole is much greater than its 20,000 or so

---

* Hood, interview with Coveney and Highfield, August 14, 2021.

genetic parts, and reveal problems: when patients ask what is wrong, they want a diagnosis of the underlying cause, not their symptoms repackaged in a confusing way.

When that day comes, you will no longer have to depend on a brief consultation, inscrutable medical records, gnomic diagnoses or incomprehensible scans. Each medical examination and scan, along with information from a plethora of smart devices, will update your avatar so you can see the site of inflammation or injury, watch how your heart beats or, as your vessels dilate, allow you to study how your blood flows faster and changes in hue from blue to red as it passes through your lungs.

Intriguing though this insider's view will prove, the ability to produce a digital twin has much broader implications. Virtual You will redefine what we mean by disease, alter what we mean by health and normality, and change the very nature of medicine itself. The point of view of medicine will shift from an enterprise obsessed with the past into one focused on the future, from patient symptoms to both symptoms and the patient's environment, from one driven by diagnoses that are often too late to be of value to predictions and even—with sufficient computer power—a glimpse of how the way we live today could alter our health in years to come. Vast resources and efforts are invested towards understanding the pathogenesis of disease. Virtual You will help shift attention from understanding the origins of disease towards understanding what some call salutogenesis, from the Latin *salus* (meaning "health") and the Greek *genesis* (meaning "origin")—in other words, how to become healthier.

Healthcasts

Just as a weather forecast relies on a mathematical model of the Earth's atmosphere, running on a supercomputer using data from meteorological stations, satellites and other instruments, one day mathematical models of a patient will be constantly updated with information from the body to produce healthcasts. There could be an average model that gives overall insights into lifestyle, symptoms and so on. More sophisticated insights can come from digital twins, primed with a person's data, from DNA code to body scans, that can be personalised.

Your digital twin would be your lifelong, personalised clone that ages just like you, as it is constantly updated with each measurement, scan or medical examination that you have at the doctor's, along with behavioural, environmental, genetic and a plethora of other data. One day in the next two decades, it will be possible to conduct a trial on a digital copy of you rather than infer your possible fate on the basis of previous trials on other people who bear a resemblance to you.

Why stop at one virtual twin? Within a supercomputer, many virtual versions of you can simultaneously explore your many possible futures, depending on your diet, medication, lifestyle and environment. Your virtual future will arrive more quickly than actual reality. With enough computational power, life can be sped up to chart the long-term effects of a novel drug on aging, quality of life and even time of death. An army of virtual twins could be born, age and then die on a supercomputer within a few hours. Pooling data between twins and by linking twins could improve predictions.[23]

In this way, digital doppelgängers of patients will herald the dawn of truly personalised medicine, where virtual drug trials are conducted on thousands of digital twins to forecast a vast range of possible futures. But, of course, we can go even further than treating disease. We can use virtual humans to show how to prevent illness as well. A virtual copy of you can be used to simulate the accumulated effects of small changes in diet, exercise and lifestyle over the decades. Just as we can predict weather with reasonable accuracy, so simulations based on many virtual copies will be able to produce healthcasts, couched in probabilities, that can give you a realistic idea of what a change in diet and lifestyle could really mean for you.

Researchers are using populations designed with demographic data to understand the social, economic and environmental factors that affect health. Imagine the possibilities if epidemiology and disease modelling could draw on the insights gleaned from studies of ensembles of digital twins, from the spread of virtual pandemics to the management of complex conditions, such as multiple sclerosis,[24] to understanding the incidence of cancer.

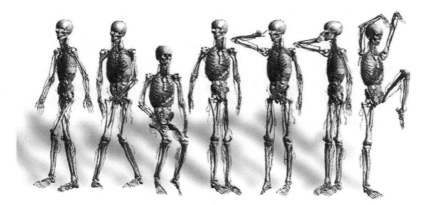

FIGURE 51. Skeletal virtual population. (IT'IS Foundation)

But, as we integrate with our digital copies and they grow more like us, other issues may emerge: What happens to our twins when we die? Should they be donated to science, or will families feel protective because, in some ineffable way, these twins contain the quintessence of their loved ones?

Beyond Normal

Many new issues will be prompted by Virtual You, some philosophical, many ethical. Like any aspect of medical data, counselling will be needed to help people make sense of healthcasts, and governance mechanisms will be needed to safeguard the rights of all those who have virtual twins. As mentioned earlier, enduring privacy concerns that were raised in the context of genomics will be even more pressing and relevant in the era of Virtual You.

But we need openness too. Given that this technology will not be available to everyone, there must also be transparency about data usage and derived benefits. Some, like the Royal Society and British Academy, have called for a set of high-level principles to shape all forms of data governance, ensuring trustworthiness and trust in the management and use of data as a whole.[25] As has already been seen with machine learning, if a group is misrepresented in the data used to build virtual humans, that group may receive suboptimal treatment that exacerbates existing racial or societal biases.[26]

Questions will arise about the degree to which you will be able to make your own decisions and to what extent they will be shaped by simulations and digital twins that propose the best approach to lifestyle and treatments. The differences between us will be made more transparent by the rise of digital twin technology, no doubt leading to yet further examples of bias against the poor, the virtual human have-nots, perhaps leading to entirely new kinds of discrimination.

The rise of big personal data and misuse of artificial intelligence could pave the way for the dystopian segmentation of society, caused by deliberate or inadvertent discrimination, something already recognised as being widespread.[27] The rise of digital twins will underline the importance of governance concerning personal data. Their constitution along with their social and legal administration will likely be based on a new alliance between our digital and biological selves, one protected by the state.[28]

We will also have to think anew about what it means to be "healthy" and "normal." Rather than rooting what these terms mean in how we look and feel right now, such definitions will come to depend on how our digital twins age, where they can produce graphic and detailed evidence of the impact of our lifestyles over the years to come. A person who is deemed healthy by today's standards could be seen as unhealthy, suffering from a "symptomless illness" or perhaps being "asymptomatically ill," if their healthcasts suggest they could live significantly longer by changing their lifestyle, adopting a new diet or using an implant or drug.[29]

A person will no longer be defined as "normal" by simply comparing them with averages of populations but by measuring what is "normal" for them. You may feel just fine when your aging digital twin reveals you to be suboptimal and, based on this healthcast, your doctor recommends you undertake treatment to make you normal so you can achieve an average life span. Modelling changes over long timescales is challenging, "but somehow capturing the irreversible chemical changes that can't be repaired and seeing how they impact on all the virtual organs and disease states is critical," in the words of Philip Luthert, of the UCL Faculty of Brain Sciences.* Comparison with other species offers other insights, according to Luthert. "It's intriguing that just as average life span of different species links to body mass and metabolic rate through a power law, genetically driven degenerations scale similarly. So, if we could convince a photoreceptor with a mutation that it was in an elephant, as opposed to a human, the rate of cell loss could be dramatically lowered whilst function was maintained."

Virtual copies of ourselves will force us to reconsider many aspects of life that we currently take for granted. We may well witness a profound change in what we consider to be healthy and diseased. What we regard as a therapy might also be subject to a rethink if a digital twin reveals the long-term implication of what today is regarded as a trivial aspect of diet or lifestyle.

Even if your digital twin might provide a vivid glimpse of your life span, along with the means to discover how it can be extended, what if we don't like what our second self is telling us? How much will the patterns of potential futures alter our identity? Could the mere fact that institutions such as insurers know your risk of being weak or sick or short-lived actually make you sick, weak or short-lived? What if we want to harm our digital oracle? With the widespread use of analogue processors, there may even come a day when we have to care about the rights of our virtual twin. However, we need to exercise a degree of modesty about this possibility, which remains distant.

By the same token, virtual twins will provide a powerful way to work out the impact of government policies on society. The implications of changes in healthcare, housing, diet and more can be safely tested and probed in virtual populations. Evidence-based policymaking is an intoxicating thought, perhaps a terrifying one for politicians, and the trustworthiness of simulations might become a battleground between different political parties, as has already been seen in the case of using modelling to guide health policies to deal with COVID-19.

Technology of any kind is a double-edged sword, offering both risks and opportunities for individuals and for society. Our ability to

---

* Philip Luthert, email to Peter Coveney, August 16, 2021.

compare populations of real and virtual people will at least help us to define normality, to establish when a person deviates from the norm as the result of disease and show exactly when treatment transforms into enhancement. However, the ability to define what is normal based on a person's detailed data is likely to result in a very different concept of "normal" than that derived from population studies.

Virtual You will further blur the boundaries between human and machine. Even with analogue technology, we doubt that virtual consciousness (whatever is really meant by that term) will wink into life any time soon. Yet virtual twins will transform the human condition by increasing confidence in methods that we can use to enhance human intellect and physiology, whether by implants, drugs or gene editing.

Your twin will offer doctors a risk-free means to test experimental treatments and allow them to explore more controversial ideas. Rather than fix or repair parts in the body, the capabilities of Virtual You could be enhanced with prosthetics, genetics and implants, for example, in an effort to extend mental performance, or to "see" beyond the ends of the visible spectrum, whether heat or ultraviolet, or to wring more energy from a given amount of food or less when it comes to those who are overweight. Anything can be tried and tested with no risk to yourself, no need for animal research, providing new insights and solutions to old problems.

As it becomes easier to augment our mind and body, just as an engineer can turbocharge a car, the concept of what it means to be human could gradually change. In this way, virtual humans will enable us to plan optimally for a posthuman future. That may sound fanciful, but we have already started down that path, given how the boundary between human and machine is already blurred by the vast range of devices and prosthetics currently in use, from cochlear implants to pacemakers and heart valves.

The rise of the digital twin could and should give many people pause for thought about where this technology is taking us. Some may welcome how digital twins allow them to take more responsibility for their destiny, others may condemn this as unnatural. We should not let the downsides deter us, however. The Janus-like nature of technology has been apparent for more than a million years: ever since we harnessed fire, we knew we could use it to stay warm and cook but also to burn down our neighbours' houses and fields.

Virtual You should be welcomed because digital twins will ultimately give us much more responsibility for our own future. When it comes to giving us control over our destiny, there's no doubt that, for the first time, people will be confronted with real choices about their health, ones that are informed by data and understanding and not based on guesswork, influencers or how people a bit like themselves responded in similar circumstances many years ago.

Medicine will shift from making sense of how past events created symptoms to using digital twins to anticipate how to influence health, well-being and longevity. The quantum pioneer Niels Bohr reputedly quipped that it is difficult to make predictions, especially about the future. When digital twins are commonplace in medicine, that will no longer hold true.

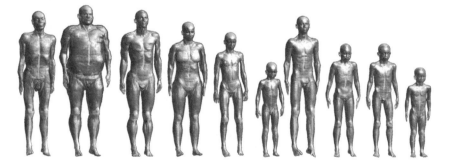

FIGURE 52. Virtual population, a set of detailed high-resolution anatomical models created from magnetic resonance image data of volunteers. (IT'IS Foundation)

Acknowledgments

We have discussed writing this book for more than a decade. Several threads of thought in these pages date back even further, and rest on our two earlier collaborations, *The Arrow of Time* and *Frontiers of Complexity*, which in turn stem from our conversations when we were both based at the University of Oxford in the seventies and eighties. We have come a long way since our very first discussions about the nature of time, computation and complex systems.

We owe an enormous debt of gratitude to the many people who have helped us along the way. And, as we hope we have made clear throughout this book, the efforts to create and develop Virtual You are by a diverse, huge and global group of scientists. We have done our best to include as many as we can, and give a sense of the vast scope of this international effort, though there are many more that we should have mentioned. To them, sincere apologies.

For the production of this book, we are enormously grateful to our agent, Peter Tallack, to Louisa Pritchard of Louia Pritchard Associates, and our Princeton University Press editor, Ingrid Gnerlich, who enlisted invaluable advice from Eric Stahlberg of the Frederick National Laboratory for Cancer Research, Maryland. Ingrid's assistant, Whitney Rauenhorst, also helped with the preparation of the final manuscript. Thanks are also due to our copyeditor, Jennifer McClain, production editor, Natalie Baan, copywriter, David Campbell, designer, Chris Ferrante, illustration manager, Dimitri Karetnikov, indexer, Julie Shawvan, publicist, Kate Farquhar-Thomson, and marketer, Julia Hall. We had a great deal of support of many kinds, from file formatting to reference and permissions wrangling from Peter's assistant, Nick Laver, at UCL. Paul Franklin and Sonia D'Orsi provided us with timely and invaluable design counsel. When we were putting together the proposal a few years ago, we also benefitted from excellent advice from Lord Alli.

Many people have offered us critiques. In particular, we would like to thank the following for constructive and detailed criticisms of various drafts of Virtual You:

Julia Brookes, Jonathan Dagley, Jane Denyer, Katie Dowler, Ross Fraser, Matthew Freeman, Heather Gething, Chris Greenwell, Sarah Harris, Phil Luthert, Eamonn Matthews, Carmen Nassé, Peter Sloot, Eric Stahlberg and Andrea Townsend-Nicholson.

Many people read and commented on parts of the book, helping us to hone clarity and presentation: Bruce Boghosian, Colleen Clancy, Christopher Coveney, Markus Covert, Omer Dushek, Timothy Elliot, Nader Engheta, Richard Evans, Steve Furber, David Goodsell, Jonathan Gorard, Gunnar von Heijne, Peter Hunter, Shantenu Jha, Viktor Jirsa, Dieter Kranzlmüller, Leslie Loew, Peter Love, Chao-Yang Lu, Paul Macklin, Jo Marchant, Alberto Marzo, Andrew Narracott, Steven Niederer, Denis Noble, Paul Nurse, Tim Palmer, John Pendry, Gernot Plank, Venki Ramakrishnan, Blanca Rodriguez, Johannes Schemmel, Koichi Takahashi, Ines Thiele, Kathryn Tunyasuvunakool, Mariano Vázquez, Shunzhou Wan, Stan Williams, Martin Winn and Paul Workman.

We collaborated to create an IMAX movie—*Virtual Humans*—with a large team that included Fernando Cucchietti and Guillermo Marin of the Barcelona Supercomputing Centre, who directed it, along with the help of Ben Lukas Boysen, Sol Bucalo, Edi Calderon, Elena Coveney, Rogeli Grima, Alfons Hoekstra, Michael Kasdorf, Dieter Kranzlmüller, Irakli Kublashvilli, Elisabeth Mayer, Paul Melis, Thomas Odaker, Carlos Tripiana, Casper Van Leeuwen, Marco Verdicchio, David Vicente, Markus Widemann and Gábor Závodszky. The movie showcases the work of a host of institutions: University of Amsterdam, ITMO University, UCL, Leibniz Supercomputing Centre, University of Copenhagen, University of Leeds, University of Sheffield, University of Oxford, Barcelona Supercomputing Centre and the Qatar Robotic Surgery Center. We also owe a debt of gratitude to Trigital, Medtronic, Alister Bates, Jorge Bazaco, Felix Bueno, Denis Doorly, Dan Einstein, Jordi Torrandell and Diana Velez. The project was supported by the EU Horizon 2020 programme.

We would also like to thank the following for consenting to interviews or for providing information: Magdalini Anastasiou, Ludovic

Autin, Michael Blinov, Bruce Boghosian, Markus Covert, Omer Dushek, Nader Engheta, Steve Furber, David Goodsell, Jonathan Gorard, Alfons Hoekstra, Leroy Hood, Peter Hunter, Shantenu Jha, John Jumper, Suhas Kumar, Seth Lloyd, Leslie Loew, Chao-Yang Lu, Jo Marchant, Martina Maritan, Karen Miga, Steve Niederer, Denis Noble, Paul Nurse, Simona Olmi, Grace Peng, Roger Penrose, Tom Ritchie, Blanca Rodriguez, Johannes Schemmel, Rick Stevens, Mariano Vázquez, Craig Venter, Marco Viceconti, Stan Williams and Paul Workman. Jon Dagley did excellent work in correcting the proofs.

Various people kindly helped us with permissions: Fred Alcober, Karen Barnes, Danielle Breen, John Briggs, Clint Davies-Taylor, Jon Fildes, Jessica Hill, Bjoern Kindler, Bryn Lloyd, Yorgo Modis, Silvia Mulligan, Liz Pryke, Cristobal Rodero, Danielle Rommel, Duncan Smith and Arwen Tapping.

Writing such a wide-ranging interdisciplinary book would be impossible without the support of our institutions and research funding. They include UCL (particularly Peter's numerous colleagues at the Centre for Computational Science, the Advanced Research Computing Centre and many more), Amsterdam (the Computational Science Lab), Amsterdam (the Computational Science Lab), Yale, Oxford, the UKRI Medical Research Council and the Science Museum Group (thanks, in particular, to Roger's assistant, Katie Dowler, her successor Laura Ambrose, and director Sir Ian Blachford), along with many colleagues and collaborators down the years, including those within the CompBioMed Centre of Excellence, VECMA and various other large international projects primarily funded by the EU and the Engineering and Physical Sciences Research Council. Roger would also like to thank the Jaipur Literature Festival, notably the Soneva Fushi edition.

We both became more productive during the writing of this book as a result of the pandemic that struck at the start of 2020. Even so, the journey to publication has been long and daunting, and we have hugely benefitted from the love, encouragement and support of our families: Julia, Holly and Rory, along with Samia, Elena and Christopher.

Appendix

Towards a Virtual Cosmos

The power of computers to simulate the workings of the human body at a range of scales, from atom to cell to organ to body, raises a wider question. How much further can we push these simulations?

Among the most demanding tests of modern supercomputers are cosmological simulations, from the dynamics of dark matter to the formation of large-scale structures, as gravity forges filaments and galaxies. Thousands of universes can be created this way.[1] However, some want to go even further, emboldened by advances in modelling the flow of blood from the "bottom up," which we described in Chapter Seven. They believe that it should be possible to re-create the fundamental physics of the cosmos from scratch.

In *Frontiers of Complexity*, we recounted the history of one way to model the universe, using what are called cellular automata that date back to work by John von Neumann and Stan Ulam, whom we encountered earlier. They consist of a collection of "coloured" cells in a grid that evolves through a number of discrete time steps according to a set of rules based on the states of neighbouring cells, hence the name *cellular automata*.

Using this approach, the British mathematician John Conway (1937–2020) came up with a simple recipe for complexity based

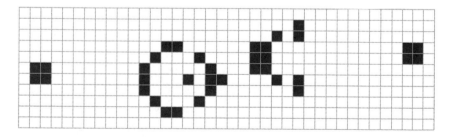

FIGURE 53. A "glider gun," a Game of Life pattern with unbounded growth. Adapted from "Gosper glider gun" (Wikimedia Commons)

on a cellular automaton that he called the Game of Life.[2] Imagine an outsize chess or checker board—a toy universe, if you will—in which each square is either black or white. The sequence of squares on one horizontal row determines the sequence of the one below it, according to a simple set of rules. From these simulations, you can generate remarkably complex, even lifelike, patterns from simple laws. That led to the idea that cellular automata could, in principle, model any real-life process.

Support for this view came from understanding the profound problem of turbulent flow. In a simulation of blood pumping through an artery, for example, the "true" continuous fluid flow is described by the Navier-Stokes equations but, as ever, the flow has to be broken up into discrete (that is, finite) slabs of space and regimented by ticks of a clock, like so many pieces of Lego, to solve these equations in a computer.

However, more than three decades ago, a group of intrepid theoretical physicists realised that they could turn this approach on its head. By defining the fluid as a cellular automaton—a regular lattice with point particles of fixed mass on it—they were able to show that the Navier-Stokes equations provide an emergent description of the collective dynamics of all the particles moving on the lattice, when the fluid properties are measured at much longer length and timescales compared to the size of the lattice and time spacings.

Among those who first discovered this was the British computer scientist Stephen Wolfram, who stepped out of a successful academic career to develop his Mathematica software, and has been working on cellular automata for decades. Visualising the world in terms of cellular automata has turned out to be a powerful method for solving fluid dynamics equations on supercomputers, not least in the form of the HemeLB software that Peter's team uses to predict the blood flow around a virtual body on emerging exascale machines.

This approach of using cellular automata on a square chessboard-like array with particles moving along edges between vertices had been tried long ago, in the 1970s, but the simulated flow ended up depending on its orientation with respect to the underlying lattice.[3,4] It is one thing to guide a flow of blood within a vein or artery, quite another to have blood that innately prefers to flow more in one arbitrary spatial direction than another. The reason comes down to the mathematical shorthand used, a multidimensional entity called a tensor, which Wolfram found only becomes indifferent to the direction of flow—that is, the flow is isotropic—if you use a hexagonal rather than a square lattice. This marked a remarkable triumph for the modelling of fluid dynamics based on extremely simple underlying rules, demonstrating the power of "complexity theory" for describing macroscopic fluid flow, including turbulence.

But there is a price to pay for this success. Multiscale modelling breaks down when mapping from the lattice to the molecular world. Yes, these discrete lattice models work well in solving the continuum fluid dynamics equations used to model blood flow. But the underlying lattice model does not correspond with how molecules actually move in continuous space, so a vast amount of information is lost in going from a molecular description in continuous space to a fictitious discrete space with only a finite number of velocities.

For many scientific applications, however, this coarse-graining is tantamount to throwing the baby out with the bathwater. While this scheme provides a great deal of important insight concerning the "universal" aspects of fluid dynamics, essentially all the chemical details of the underlying molecules are simply discarded (albeit attempts are frequently made to retrofit these details). There is something decidedly disconcerting about seeking to replace the continuous and powerful symmetries we encounter in particle physics and the behaviour of molecules with an artificial "Lego-land," one that accurately provides some emergent properties in fluid mechanics but ignores the realities of the microscopic world.

Notwithstanding these limitations, Wolfram is among those who find the computing metaphor so heady that he is convinced that at the "deepest level," at the smallest of all time and space scales, nature runs on ultrasimple computational rules based on discrete mathematics. While the classical view of the world is analogue, he thinks there is an underlying discreteness to the structure of the universe (however, overall, there is no agreement among physicists on whether space-time remains smooth and continuous at the shortest distance scales, or if it becomes coarse and grainy[5,6]). He replaced the cells of a cellular automaton lattice with a generalisation, namely, a graph, a network of points with connections between them, that grows and morphs as the digital logic of the cosmic code advances, one step at a time (graphs are necessary because lattices are too inflexible to encompass the laws of the universe).

Wolfram's argument goes that, just as the continuum equations of fluid mechanics emerge from a specially constructed, spatially discrete lattice model, for which basic properties, such as mass and momentum, are conserved, so the continuum of what we perceive as reality emerges from a small set of rules and graphs. In his recent book, *A Project to Find the Fundamental Theory of Physics*, he writes about his "strong belief that in the end it will turn out that every detail of our universe does indeed follow rules that can be represented by a very simple program—and that everything we see will ultimately emerge just from running this program"[7] (although some of the results that arise from it may be uncomputable, as discussed in Chapter Two).

Wolfram and his collaborator Jonathan Gorard, a mathematician at the University of Cambridge and a consultant at Wolfram Research, found that they could recapitulate aspects of quantum theory and Einstein's general theory of relativity, the two pillars of modern physics.[8,9] "I have been surprised by how much we have been able to reproduce," Gorard told us. "But there are an awful lot of mathematical techniques that are having to be developed." Rather than do experiments and then drill down to reveal fundamental theory, the traditional approach used for more than three centuries, Gorard and Wolfram are excited by "inverting the paradigm" and compute the consequences of bottom-up models to see if they match experimental observations.*

It is no accident, Gorard adds, that these models "reflect ambient technology": the French philosopher René Descartes (1596–1650) had a mechanical worldview in the era of mechanical timekeeping; when Newton and Leibniz created the mathematics of change, calculus, that then underpinned mechanistic explanations of nature based on a view of time and space that was analogue and continuous. In a similar way, the discrete mathematical worldview of Wolfram has emerged in the era when digital computers are part of everyday life.

---

* Jonathan Gorard, interview with Peter Coveney and Roger Highfield, October 1, 2021.

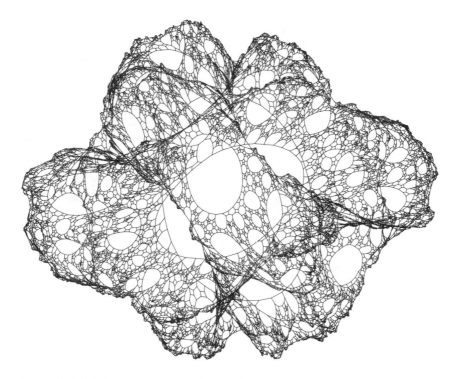

FIGURE 54. Building general relativity from the bottom up. (© Stephen Wolfram, LLC)

There is something refreshing about taking a fundamentally different approach to physics. While the lattices they used in fluid dynamics do not map onto the real world, the graininess of their proposed models is constructed from graphs that are smaller than the tiniest measurable features of nature at the Planck scale, below which physics as we understand it today breaks down. When we say tiny, that is overstating it: we mean less than time intervals of around 10^{-43} s and length scales of around 10^{-35} m.

The acid test of the validity of their approach will come if and when their model tells us something new about the universe. As we write, they are seeking to make predictions, such as about the final stages of the merging of black holes when their discrete computational view delivers different outcomes from existing models; and that the universe might have spatial pockets that are slightly more or less than three-dimensional, which may be detectable by their effects on galaxy formation, the passage of neutrinos through matter; and more speculative ideas about other superlight particles.

We should put this work in its proper, cosmic, context. We only really understand around 5% of our surroundings—the visible universe, from our bodies to our Sun, stars and galaxies, which are made of ordinary or baryonic matter—while the rest appears to be made of a mysterious, invisible substance called dark matter (about 25%) and an equally unfathomable repulsive force that behaves like antigravity known as dark energy (about 70%). Given our ignorance about the universe, there is plenty of room for new thinking.

Glossary

Defined terms are shown in **bold**.

Algorithm: A stepwise procedure which can be performed mechanically, and hence implemented on a machine. In computer science, it describes a **program**, the logical sequence of operations carried out by software.

Algorithmic complexity: A measure of the complexity of a problem given by the size of the smallest **program** which computes it or a complete description of it. Simpler things require smaller programs.

Amino acids: The molecular building blocks of **proteins**.

Analogue: Continuously variable, like the movement of mercury within a thermometer.

Artery: Vessel that carries blood away from the heart.

Artificial intelligence: A branch of computer science concerned with the design of computers that have attributes associated with intelligence, such as learning and reasoning. There are many kinds of AI, such as **artificial neural networks**. **Machine learning** is a term that is frequently used synonymously with AI.

Artificial neural network: Also called a neural net, a computer program or model that is loosely inspired by the structure of the brain, consisting of interconnected processing units, often arranged into layers. When the network has more than three layers, the term *deep neural network* is used.

Atoms: The building blocks of all the molecules in living things. Around a trillion carbon atoms or water molecules would fit on the period (full-stop) at the end of this sentence.

Atrium: One of the two chambers (left and right) in the upper part of the heart.

Attractor: A way to describe the long-term behaviour of a system, its destination, or end state. Equilibrium and steady states correspond to fixed-point attractors, periodic states to **limit-cycle attractors** and chaotic states to **strange attractors**.

Axon: The long fibre extending from a **neuron**—nerve cell—that carries a signal to other neurons.

Base: The molecular units—"letters" of genetic code, called nucleotides—that provide the variation in a strand of DNA or RNA that carries information. DNA has four bases—thymine, cytosine, adenine, and guanine—while RNA contains uracil instead of thymine.

Belousov-Zhabotinsky reaction: A chemical reaction which displays a remarkable wealth of self-organising features, manifested as patterns in space and time.

Bit: Abbreviated form of "binary digit." A bit is the smallest unit of information in a binary number system. The value of a bit is usually referred to as a 1 or 0.

Bottom up: An approach to a problem that relies on extracting information directly from all its myriad details as opposed to the top-down approach, which uses rules, models and knowledge.

Byte: A group of eight binary digits.

Capacitor: A device for storing an electrical charge.

Cardiovascular system: Includes the heart, blood and a vast network of blood vessels. Also called the circulatory system.

Cell: A discrete, membrane-bound portion of living matter, the smallest unit capable of an independent existence. Forty human skin cells placed in a line would add up to around a millimetre.

Cellular automata: A spatially and temporally discrete dynamical system consisting of cells separated by lines and vertices. At each time step, a simple **algorithm** controls how the state of each cell or vertex changes, based on the state of neighbouring cells or vertices.

Central processing unit (CPU): The part of a **computer** which performs the actual execution of instructions.

Cerebral cortex: The outer (grey matter) layer of the cerebral hemispheres, evolution's most recent addition to the nervous system.

Chaos: Apparently random behaviour—in fact, there is deep order—which is exquisitely sensitive to initial conditions. The branch of mathematics that is used to deal with such behaviour is called chaos theory.

Chromosome: A long strand of **DNA**, containing hundreds to thousands of genes, in a protective package. There are 23 pairs in all human cells, except eggs and sperm.

Complexity: The study of the behaviour of collections of units which are endowed with the potential to evolve in time, culminating in emergent properties that are more than the sum of their parts.

Computable number: A number that can be computed by an individual **Turing machine**.

Computation: A calculation (usually of numbers) performed by means of an **algorithm**.

Computer: A device which operates on data (input) according to specified instructions (**program**, or **algorithm**). In a digital computer, all values are represented by discrete signals—such as 1s and 0s—rather than by continuously variable values as found in an **analogue** device.

Computer-aided tomography: A CT (or CAT) scanner works by sending multiple X-ray beams through the body at varying angles. This is called tomography. Detectors record how the beams pass through sections of the body and the data are processed by a computer into virtual "slices," which are then reconstructed into 3D images.

Concentration gradient: Change in the concentration of a substance from one volume to another.

Cortex: From the Latin word for "bark," the thin rind on the hemispheres of the brain where most of the information processing takes place. The cortex is grey—hence "grey matter"—because nerves in this area lack the insulation that makes other parts of the brain appear white.

Determinism: The theory that a given set of circumstances inevitably produces the same consequences. More mathematically, given the initial conditions of a set of equations describing the time evolution of a system, its future behaviour will be exactly known.

Differential equations: Equations involving the instantaneous rate of change of certain quantities with respect to others, most usually time and space. Newton's equations of motion, for example, are differential equations linking the force experienced by a body to the instantaneous rate of change of its velocity with time.

Dissipative structure: An organised state of matter when a system is maintained far from thermodynamic **equilibrium**. Examples include the patterns formed by the **Belousov-Zhabotinsky** chemical reaction.

DNA: Deoxyribonucleic acid, the vehicle of inheritance for most creatures. The complex giant nucleic acid molecule carries instructions to make **proteins**, the basic building blocks of life.

Dynamical systems: General term for systems whose properties change with time. Dynamical systems can be divided into two kinds, conservative and **dissipative**. In the former, the time evolution is reversible, in the latter it is irreversible.

Emergent property: A global property of a system that consists of many units. For example, consciousness is an emergent property of the many **neurons** in the brain.

Enzyme: A biological catalyst, usually comprising a large protein molecule which accelerates essential chemical reactions in living cells.

Entropy: A quantity which determines a system's capacity to evolve irreversibly in time. Loosely speaking, we may also think of entropy as measuring the degree of randomness or disorder in a system.

Epigenetics: Changes in how genes are used in the body that take place without changes to the genes themselves, for instance, during development.

Epilepsy: A disorder of brain function characterised by sporadic recurrence of seizures caused by an avalanche of discharges of large numbers of neurons.

Equilibrium: A state of time evolution at which all capacity for change is spent. Equilibrium thermodynamics is concerned exclusively with the properties of such static states.

Evolution: From the Latin *evolutio*, "unfolding." The idea of shared descent of all creatures, from yeast to humans, credited to Charles Robert Darwin (1809–1882) and Alfred Russel Wallace (1823–1913). Diversity can arise through mutation, random changes in **DNA**, from generation to generation.

Exascale: Computing systems capable of a million million million (10^{18}) floating-point operations, or **flops**.

Feedback: The general principle whereby the results produced in an ongoing reaction become factors in modifying or changing that reaction.

Fitness landscape: A landscape representing the fitness measures or cost functions of a problem, whether travelling salesmen, spin glasses or the reproductive capability of a monkey.

Flops: Floating-point operations per second in a computer, that is, an operation where numbers are represented by a floating-point number, used since the 1950s to approximate real numbers on digital computers.

Fluid mechanics: The study of the behaviour of fluids (liquids and gases) at rest and in motion.

Fractal geometry: From the Latin *fractus*, meaning "broken." The geometry used to describe irregular patterns. Fractals display the characteristic of self-similarity, an unending series of motifs within motifs repeated at all length scales.

Gene: A unit of heredity comprising chemical **DNA**, responsible for passing on specific characteristics from parents to offspring.

Gene expression: Genes that are used in the body are said to be "expressed." Different cell types express different sets of genes.

Genetic algorithm: A method to find a good solution to a problem that borrows ideas from evolution.

Genetic code: The sequence of chemical building blocks of DNA (bases) which spells out the instructions to make amino acids, the molecular building blocks of proteins.

Genotype: The genetic constitution of an individual.

Geometry: The branch of mathematics concerned with the properties of space.

Giga: Prefix signifying multiplication by 1,000,000,000.

Graphics processing unit: A GPU is a chip or processor originally designed to accelerate the display of graphics on an electronic device.

Hardware: The physical parts of a **computer** consisting of mechanical, electrical and electronic components.

Integrated circuit: The "silicon chip," a small piece of semiconductor material that is fabricated so that it contains a complete electronic circuit.

Intractable: A problem that is so demanding that the time required to solve it rapidly spirals out of control. Many of the most challenging optimisation problems in science and engineering, from biology and drug discovery to routing and scheduling, are known as **NP-complete**, where the number of operations that must be performed in order to find the solution grows exponentially with the problem size.

Irrational number: A number represented by a nonterminating aperiodic sequence of digits that cannot be expressed as an exact fraction. One example is π.

Irreversibility: The one-way time evolution of a system, giving rise to an arrow of time.

Knowledge-based system: Computer program that uses an encoding of human knowledge to help solve problems.

Life: Defined in many different ways, including as a self-sustained chemical system capable of undergoing Darwinian **evolution**.

Limit-cycle attractor: An **attractor** describing regular (periodic or quasi-periodic) temporal behaviour, for instance, in a chemical clock undergoing regular colour changes.

Linear equation: A relationship between two variables that, when plotted on Cartesian axes, produces a straight-line graph.

Machine code: The sequence of binary commands which is executed by the hardware of a **computer**.

Machine learning: An inference-based approach to finding relationships between input and output data within digital computers by an algorithmic process known as "training" in which correlations are detected and honed through iterative adjustment of large numbers of free parameters. Often used interchangeably with **artificial intelligence**. Examples par excellence are **artificial neural networks**. Deep learning is a term used to describe such learning in artificial neural networks that have more than three layers.

Macroscopic: The level of description of phenomena which are directly accessible to our senses.

Mathematics: The science of spatial and numerical relationships.

Measurement problem: In **quantum mechanics**, the problem of accounting for the outcome of experimental measurements on **microscopic** systems.

Mega: Prefix denoting multiplication by a million.

Memory: In computer science, the space within a computer where information is stored while being actively processed.

Memristor: The fourth circuit element, along with the resistor, capacitor and inductor. A memristor has a resistance that "remembers" what value it had when the current was last turned on.

Microscopic: A term used to describe tiny dimensions, compared with the everyday or **macroscopic** dimensions of the world that can be directly perceived with the senses.

Microtubule: Tiny tube found in almost all cells with a nucleus. Microtubules act as scaffolding to help define the shape of a cell.

Molecular biology: The study of the molecular basis of life, including the biochemistry of molecules such as **DNA** and **RNA**.

Morphogenesis: The evolution of form in plants and animals, for example.

MRI: Magnetic resonance imaging uses magnetic fields, radio waves and a **computer** to generate images of the inside of the body. Unlike X-rays, an MRI scan can visualise soft tissue, such as the organs and blood vessels.

Mutation: A change in **genes** produced by a change in the **DNA** that makes up the hereditary material of an organism.

Natural selection: The evolutionary process whereby gene frequencies in a population change through certain individuals producing more descendants than others because they are better able to survive and reproduce in their environment. The accumulated effect of natural selection is to produce adaptations.

Neural network: The brain's actual interconnected mesh of **neurons**. Neural networks may also be electronic, optical or simulated by computer software.

Neuromorophic computing: A branch of computing that mimics structures in the brain and nervous system.

Neuron: The nerve cell that is the fundamental signalling unit of the nervous system.

Neurotransmitter: A chemical that diffuses across a **synapse** and thus transmits impulses between nerve cells.

Nonequilibrium: The state of a **macroscopic** system which has not attained thermodynamic **equilibrium** and thus still has the capacity to change with time.

Nonlinear: Behaviour typical of the real world, meaning in a qualitative sense "getting more than you bargained for," unlike linear systems, which produce no surprises. For example, **dissipative** nonlinear systems are capable of exhibiting **self-organisation** and **chaos**.

NP-complete: The hardest type of **NP problem**.

NP problem: When the time required to solve a problem increases in an exponential fashion (something to the power of N, where N measures the size of the problem). These problems are called **intractable** because the time required to solve them rapidly spirals out of control. Even the raw power of a computer has little effect. These problems, which are not solvable in polynomial time but are bounded by an exponential power of N, are said to be in the class NP.

Nucleic acid: "Letters" of the genetic code present in all life-forms on Earth. The two types, known as **DNA** and **RNA**, form the basis of heredity. The nucleotides in DNA are composed of a base and other chemicals (deoxyribose and at least one phosphate group). The bases are adenine (A), cytosine (C), guanine (G), and thymine (T). In RNA, the base uracil (U) takes the place of thymine.

Number theory: The abstract study of the structure of number systems and the properties of positive integers (whole numbers). The process used to find the best solution to a problem.

Optimisation: Method to search a landscape of possibilities for the best solution to the problem, as expressed by the lowest point on that landscape.

Ordinary differential equations (ODEs): Mathematical expressions of continuous changes in a single variable, typically one dimension in space or in time, which are used to chart the motions of the heavens in the form of celestial mechanics, rates of chemical reactions in the body, the rise of infection, stock trends and so on.

Organelles: Structures found within cells that are analogous to organs in a body.

Organic chemistry: The branch of chemistry that deals with carbon compounds.

Parallelism: Parallel processing is achieved when a **computer** is able to carry out more than one computation at the same time by using

several different **central processing units**. Concurrent processing (time-sharing) is achieved on serial machines by making the
single CPU switch between several tasks.

Partial differential equations (**PDEs**): Mathematical expressions
for how things change as a function of more than one variable,
such as time and space, the latter being capable of description
in several dimensions (often but not always three). PDEs can be
found in all our theories of the very big and the very small—such
as general relativity, electromagnetism and quantum mechanics—
and capture the way we understand heat, diffusion, sound, electromagnetism, fluid dynamics and elasticity.

Petascale: The prefix *peta* indicates multiplication by one quadril-
lion, or 1,000,000,000,000,000 (10^{15}).

Phase change: A change in the physical state of a material, such as
from solid to liquid, liquid to gas, or solid to gas. Many other more
subtle phase changes exist in various kinds of matter.

Phenotype: The characters of an organism due to the interaction of
its **genotype** with the environment.

Pixel: Shorthand for "picture element," a single dot on a computer
screen.

Polymer: A chemical substance formed by the chain-like linkage of
many simpler molecules.

Program: An **algorithm**, or set of instructions, that executes on a
computer.

Protein: A class of large molecules that are the building blocks of
living organisms, consisting of strings of amino acids folded into
complex but well-defined three-dimensional structures. Examples include hormones, enzymes and antibodies.

Quantum mechanics: The mechanics that rules the microscopic
world of subatomic particles, where energy changes occur in
abrupt, tiny quantum jumps.

Rational number: Any number that can be expressed as an exact
fraction.

Reaction: In chemistry, the coming together of atoms or molecules
with the result that a chemical change takes place, rearranging
the molecular structure so that different substances result.

Real number: Any of the **rational** or **irrational** numbers.

Reductionism: A doctrine according to which complex phenomena can be explained in terms of something simpler. In particular, atomistic reductionism contends that **macroscopic** phenomena can be explained in terms of the properties of atoms and molecules.

Reynolds number: A number used in **fluid mechanics** to determine whether a fluid flow in a particular situation will be smooth or turbulent.

RISC: Acronym for reduced instruction set computer. A microprocessor that carries out fewer instructions than traditional microprocessors, referred to as complete instruction set devices, so that it can work more quickly.

RNA (ribonucleic acid): The genetic material used to translate **DNA** into **proteins**. In some organisms, it can also be the principal genetic material.

Scalar: A quantity with magnitude but no direction. Examples include mass and temperature.

Search space: The variation in a cost function is best envisaged as an undulating landscape of potential solutions where the height of each feature is a measure of its cost.

Self-organisation: The spontaneous emergence of structural organisation on a **macroscopic** level owing to collective interactions between a large number of simple, usually microscopic, objects.

Simulated annealing: A computational method for locating states close to global minima on a (hyper) surface, commencing at high temperatures and progressively reducing the temperature following a well-defined prescription. In this way, trapping in local minima is avoided. The name is drawn from the process of annealing, whereby a material is initially heated, followed by slow cooling to boost ductility and strength, allowing the component atoms of the metal to settle into a lower energy arrangement.

Software: The programs executed by a computer system, as distinct from the physical hardware.

Statistical mechanics: The discipline which attempts to express the properties of **macroscopic** systems, say, the temperature of a gas, in terms of their atomic and molecular constituents.

Steady state: A nonequilibrium state which does not change with time.

Strange attractor: An **attractor** which has a fractal (fractional) dimension: describes chaotic dynamics in **dissipative** dynamical systems.

Supercomputer: The fastest, most powerful type of **computer**.

Surrogate models: Use of machine learning to emulate a simulation which would be too computationally costly to carry out.

Synapse: The junction between two nerve cells across which a nerve impulse is transmitted.

Terascale: The prefix *tera* indicates one million million (10^{12}).

Thermodynamics: The science of heat and work.

Transistor: A solid-state electronic component made of a semiconductor with three or more electrodes that can regulate a current passing through it. Used as an amplifier, oscillator or switch.

Turing machine: A computer program. A computer is the equivalent of a universal Turing machine.

Ultrasound: Diagnostic imaging technology that uses high-frequency sound waves—well beyond the range of human hearing—to produce pictures of the inside of the body.

Uncertainty principle: The quantum mechanical principle that says it is meaningless to speak of a particle's position, momentum or other parameters, except as the result of measurements. It gives a theoretical limit to the precision with which a particle's momentum and position can be measured simultaneously: the more accurately one is determined, the more uncertainty there is in the other.

Vector: Any quantity with a magnitude and a direction. Examples are velocity and acceleration.

Vein: Blood vessel that returns blood to the heart.

Ventricle: One of the two chambers (left and right) in the lower part of the heart.

Virus: In biology, a length of genetic material (either **DNA** or **RNA**), usually wrapped in a protein overcoat, that needs a host cell to reproduce. In computers, a short stretch of code that can copy itself into one or more larger "host" computer programs when it is activated.

Wave function: The central quantity in quantum theory which is used to calculate the probability of an event occurring, for instance, an atom emitting a photon, when a measurement is made.

Zettascale: Hypothetical future computer system one thousand times more powerful than **exascale** devices (capable of performing 10^{21} floating-point operations per second).

References

Introduction

1. YouTube. The virtual human project (posted March 12, 2018). https://www.youtube.com/watch?v=1ZrAaDsfBYY.

2. Grieves, M. & Vickers, J. Digital twin: Mitigating unpredictable, undesirable emergent behavior in complex systems. In *Transdisciplinary Perspectives on Complex Systems* (eds. Kahlen, J., Flumerfelt, S. & Alves, A.), 85–113 (Springer, 2017).

3. Shafto, M. et al. DRAFT Modeling, Simulation, Information Technology & Processing Roadmap, Technology Area 11 (NASA, 2010).

4. Negri, E., Fumagalli, L. & Macchi, M. A review of the roles of digital twin in CPS-based production systems. *Procedia Manuf.* **11**, 939–948 (2017).

5. Niederer, S. A., Sacks, M. S., Girolami, M. & Willcox, K. Scaling digital twins from the artisanal to the industrial. *Nat. Comput. Sci.* **1**, 313–320 (2021).

6. El Saddik, A. Digital twins: The convergence of multimedia technologies. *IEEE Multimed.* (2018). https://doi.org/10.1109/MMUL.2018.023121167.

7. Bauer, P., Thorpe, A. & Brunet, G. The quiet revolution of numerical weather prediction. *Nature* **525**, 47–55 (2015).

8. Abbe, C. The needs of meteorology. *Science* **1**(7), 181–182 (1895).

9. Alley, R. B., Emanuel, K. A. & Zhang, F. Advances in weather prediction. *Science* **363**, 342–344 (2019).

10. European Commission. Shaping Europe's digital future: Destination Earth. https://digital-strategy.ec.europa.eu/en/policies/destination-earth (accessed May 29, 2022).

11. Mitchell, H. H., Hamilton, T. S., Steggerda, F. R. & Bean, H. W. The chemical composition of the adult human body and its bearing on the biochemistry of growth. *J. Biol. Chem.* **158**, 625–637 (1945).

12. Proctor, J., Hsiang, S., Burney, J., Burke, M. & Schlenker, W. Estimating global agricultural effects of geoengineering using volcanic eruptions. *Nature* (2018). https://doi.org/10.1038/s41586-018-0417-3.

13. Dallas, V. & Vassilicos, J. C. Rapid growth of cloud droplets by turbulence. *Phys. Rev. E–Stat. Nonlinear, Soft Matter Phys.* (2011). https://doi.org/10.1103/PhysRevE.84.046315.

14. Morton, O. *The Planet Remade: How Geoengineering Could Change the World* (Granta, 2015).

15. Auffray, C. & Noble, D. Origins of systems biology in William Harvey's masterpiece on the movement of the heart and the blood in animals. *Int. J. Mol. Sci.* (2009). https://doi.org/10.3390/ijms10041658.

16. Noble, D. Claude Bernard, the first systems biologist, and the future of physiology. *Exp. Physiol.* (2008). https://doi.org/10.1113/expphysiol.2007.038695.

17. Nosil, P. et al. Natural selection and the predictability of evolution in *Timema* stick insects. *Science* (2018). https://doi.org/10.1126/science.aap9125.

18. Nurse, P. Biology must generate ideas as well as data. *Nature* **597**, 305 (2021).

19. Charlton, W. Greek philosophy and the concept of an academic discipline. *Hist. Polit. Thought* **6**, 47–61 (1985).

20. Coveney, P. & Highfield, R. *The Arrow of Time: A Voyage through Science to Solve Time's Greatest Mystery* (W. H. Allen, 1991).
21. Coveney, P. & Highfield, R. *Frontiers of Complexity: The Search for Order in a Chaotic World* (Faber, 1996).
22. Hunter, P., Robbins, P. & Noble, D. The IUPS human physiome project. *Pflugers Archiv Eur. J. Physiol.* (2002). https://doi.org/10.1007/s00424-002-0890-1.
23. Hernandez-Boussard, T. et al. Digital twins for predictive oncology will be a paradigm shift for precision cancer care. *Nat. Med.* **27**, 2065–2066 (2021).
24. Hunter, P. et al. A vision and strategy for the virtual physiological human in 2010 and beyond. *Philos. Trans. R. Soc. A Math. Phys. Eng. Sci.* (2010). https://doi.org/10.1098/rsta .2010.0048.
25. Human Brain Project. *The Human Brain Project: A Report to the European Commission.* HBP-PS Consortium (2012).
26. Kitano, H. Grand challenges in systems physiology. *Front. Physiol.* **1**, 3 (2010).

Chapter 1: The Measure of You

1. Borges, J. L. On exactitude in science. *Los Anales de Buenos Aires* **1**, 3 (March 1946).
2. Matzeu, G. et al. Large-scale patterning of reactive surfaces for wearable and environmentally deployable sensors. *Adv. Mater.* **32**, 2001258 (2020).
3. Coveney, P. V, Groen, D. & Hoekstra, A. G. Reliability and reproducibility in computational science: Implementing validation, verification and uncertainty quantification in silico. *Philos. Trans. R. Soc. A Math. Phys. Eng. Sci.* **379**, 20200409 (2021).
4. Bianconi, E. et al. An estimation of the number of cells in the human body. *Ann. Hum. Biol.* (2013). https://doi.org/10.3109/03014460.2013.807878.
5. Chellan, P. & Sadler, P. J. The elements of life and medicines. *Philos. Trans. R. Soc. A Math. Phys. Eng. Sci.* (2015). https://doi.org/10.1098/rsta.2014.0182.
6. Succi, S. *Sailing the Ocean of Complexity: Lessons from the Physics-Biology Frontier* (Oxford University Press, 2022).
7. Nowak, M. A. & Highfield, R. *Supercooperators: Altruism, Evolution, and Why We Need Each Other to Succeed* (Canongate, 2011).
8. Toker, D., Sommer, F. T. & D'Esposito, M. A simple method for detecting chaos in nature. *Commun. Biol.* (2020). https://doi.org10.1038/s42003-019-0715-9.
9. Brenner, S. Life sentences: Detective Rummage investigates. *Genome Biol.* (2002). https://doi.org/10.1186/gb-2002-3-9-comment1013.
10. Yip, K. M., Fischer, N., Paknia, E., Chari, A. & Stark, H. Breaking the next cryo-EM resolution barrier—atomic resolution determination of proteins! *bioRxiv* (2020). https:// doi.org/10.1101/2020.05.21.106740.
11. Venter, J. C. *Life at the Speed of Light: From the Double Helix to the Dawn of Digital Life* (Viking, 2013).
12. Bentley, D. R. et al. Accurate whole human genome sequencing using reversible terminator chemistry. *Nature* **456**, 53–59 (2008).
13. Branton, D. et al. The potential and challenges of nanopore sequencing. *Nat. Biotechnol.* **26**(10),1146–1153 (2008).
14. Nurk, S. et al. The complete sequence of a human genome. *bioRxiv* (2021). https://doi .org/10.1101/2021.05.26.445798.
15. Nurse, P. *What Is Life?: Five Great Ideas in Biology.* (Scribe, 2020).
16. Venter, C., interview with Peter Coveney and Roger Highfield, December 29, 2021.

17. Levy, S. et al. The diploid genome sequence of an individual human. *PLoS Biol.* **5**, 2113–2144 (2007).
18. Venter, J. C. *A Life Decoded: My Genome, My Life* (Viking, 2007).
19. Highfield, R. What's wrong with Craig Venter? *Mosaic* (2016). https://mosaicscience.com/story/craig-venter-genomics-personalised-medicine/.
20. Perkins, B. A. et al. Precision medicine screening using whole-genome sequencing and advanced imaging to identify disease risk in adults. *Proc. Natl. Acad. Sci. U. S. A.* **115**, 3686–3691 (2018).
21. Gates, A. J., Gysi, D. M., Kellis, M. & Barabási, A.-L. A wealth of discovery built on the Human Genome Project—by the numbers. *Nature* **590**, 212–215 (2021).
22. Munro, S., Freeman, M., Rocha, J., Jayaram, S. A., Stevens, T., et al. Functional unknomics: closing the knowledge gap to accelerate biomedical research. Preprint. https://www.biorxiv.org/content/10.1101/2022.06.28.497983v1.
23. Abascal, F. et al. Expanded encyclopaedias of DNA elements in the human and mouse genomes. *Nature* (2020). https://doi.org/10.1038/s41586-020-2493-4.
24. Quinodoz, S. A. et al. Higher-order inter-chromosomal hubs shape 3D genome organization in the nucleus. *Cell* **174**, 744–757.e24 (2018).
25. Sey, N.Y.A. et al. A computational tool (H-MAGMA) for improved prediction of brain-disorder risk genes by incorporating brain chromatin interaction profiles. *Nat. Neurosci.* (2020). https://doi.org/10.1038/s41593-020-0603-0.
26. Herder, C. & Roden, M. Genetics of type 2 diabetes: Pathophysiologic and clinical relevance. *Eur. J. Clin. Invest.* **41**, 679–692 (2011).
27. Ponomarenko, E. A. et al. The size of the human proteome: The width and depth. *Int. J. Analyt. Chem.* (2016). https://doi.org/10.1155/2016/7436849.
28. Venter, J. C. *Life at the Speed of Light: From the Double Helix to the Dawn of Digital Life* (Viking, 2013).
29. Highfield, R. Ultimate molecular machine plays key role in superbug fight. *Science Museum Blog* (2016). https://blog.sciencemuseum.org.uk/ultimate-molecular-machine-plays-key-role-in-superbug-fight/.
30. Farías-Rico, J. A., Selin, F. R., Myronidi, I., Frühauf, M. & Von Heijne, G. Effects of protein size, thermodynamic stability, and net charge on cotranslational folding on the ribosome. *Proc. Natl. Acad. Sci. U. S. A.* (2018). https://doi.org/10.1073/pnas.1812756115.
31. Nilsson, O. B. et al. Cotranslational protein folding inside the ribosome exit tunnel. *Cell Rep.* (2015). https://doi.org/10.1016/j.celrep.2015.07.065.
32. Highfield, R. The anatomy of a hangover. *Daily Telegraph* (January 3, 2003). https://www.telegraph.co.uk/technology/3304259/The-anatomy-of-a-hangover.html.
33. Highfield, R. The true meaning of the morning after. *Daily Telegraph* (January 3, 2003). https://www.telegraph.co.uk/technology/3304279/The-true-meaning-of-the-morning-after.html.
34. Huckvale, K., Venkatesh, S. & Christensen, H. Toward clinical digital phenotyping: A timely opportunity to consider purpose, quality, and safety. *npj Digit. Med.* (2019). https://doi.org/10.1038/s41746-019-0166-1.
35. Service, R. DNA could store all of the world's data in one room. *Science* (2017). https://doi.org/10.1126/science.aal0852.
36. Wooley, J. C. & Lin, H. S. *Catalyzing Inquiry at the Interface of Computing and Biology* (National Academies Press, 2005).
37. Campbell, E. G. et al. Data withholding in academic genetics: Evidence from a national survey. *J. Am. Med. Assoc.* (2002). https://doi.org/10.1001/jama.287.4.473.

38. Wilkinson, M. D. et al. The FAIR Guiding Principles for scientific data management and stewardship. *Sci. Data* **3**, 160018 (2016).

39. Royal Society. Data management and use: Governance in the 21st century. Joint report by the British Academy and the Royal Society (June 2017). https://royalsociety.org /topics-policy/projects/data-governance/.

40. Nurse, P. Biology must generate ideas as well as data. *Nature* **597**, 305 (2021).

Chapter 2: Beyond Bacon's Ants, Spiders and Bees

1. Bacon, F. *The New Organon, or True Directions Concerning the Interpretation of Nature* (1620). Trans. Spedding, J., Ellis, R. E. & Heath, D. D. (1863). http://intersci.ss.uci.edu /wiki/eBooks/BOOKS/Bacon/Novum%20Organum%20Bacon.pdf.

2. Steele, J. M. Babylonian observational and predictive astronomy. In *Handbook of Archaeoastronomy and Ethnoastronomy* (ed. Ruggles, C.), 1855–1862 (Springer, 2015).

3. Coveney, P. V., Dougherty, E. R. & Highfield, R. R. Big data need big theory too. *Philos. Trans. R. Soc. A Math. Phys. Eng. Sci.* (2016). https://doi.org/10.1098/rsta.2016.0153.

4. Highfield, R. & Carter, P. *The Private Lives of Albert Einstein* (Faber, 1993).

5. Highfield, R. Heroes of science. Royal Society (2012). https://royalsociety.org/science -events-and-lectures/2012/heroes-of-science/.

6. Newton, I. *Philosophiae Naturalis Principia Mathematica* (1687). https://doi.org/10.5479 /sil.52126.39088015628399.

7. Highfield, R. Interview with Stephen Hawking. *Daily Telegraph* (October 18, 2001). https:// www.telegraph.co.uk/news/science/science-news/4766816/Interview-with-Stephen -Hawking.html.

8. Thompson, S. P. *Calculus Made Easy*. 2nd ed. (MacMillan, 1914). https://www.gutenberg .org/ebooks/33283.

9. Zeeman, E. Differential equations for the heartbeat and nerve impulse. In *Biological Processes in Living Systems* (ed. Waddington, C. H.), 8–67 (Routledge, 2017).

10. de Langhe, B., Puntoni, S. & Larrick, R. Linear thinking in a nonlinear world. *Harvard Bus. Rev.* **95**(3), 130–139 (2017).

11. Cooper, N. G. & Lax, P. From cardinals to chaos: Reflections on the life and legacy of Stanislaw Ulam. *Phys. Today* (1989). https://doi.org/10.1063/1.2811052.

12. Lee, W. Y., Dawes, W. N. & Coull, J. D. The required aerodynamic simulation fidelity to usefully support a gas turbine digital twin for manufacturing. *J. Glob. Power Propuls. Soc.* **5**, 15–27 (2021).

13. Sun, S. & Zhang, T. A 6M digital twin for modeling and simulation in subsurface reservoirs. *Adv. Geo-Energy Res.* **4**, 349–351 (2020).

14. Sherwin, S. J., Formaggia, L., Peiró, J. & Franke, V. Computational modelling of 1D blood flow with variable mechanical properties and its application to the simulation of wave propagation in the human arterial system. *Int. J. Numer. Methods Fluids* **43**, 673–700 (2003).

15. Luo, Y. (罗扬), Xiao, Q. (肖清), Zhu, Q. (朱强) & Pan, G. (潘光). Pulsed-jet propulsion of a squid-inspired swimmer at high Reynolds number. *Phys. Fluids* **32**, 111901 (2020).

16. Scully, T. Neuroscience: The great squid hunt. *Nature* (2008). https://doi.org/10.1038 /454934a.

17. Fasano, C. et al. Neuronal conduction of excitation without action potentials based on ceramide production. *PLoS One* (2007). https://doi.org/10.1371/journal.pone.0000612.

18. Häusser, M. The Hodgkin-Huxley theory of the action potential. *Nat. Neurosci.* (2000). https://doi.org/10.1038/81426.

19. Barrow, J. D. *Pi in the Sky: Counting, Thinking and Being* (Clarendon Press, 1992).
20. Turing, A. M. On computable numbers, with an application to the Entscheidungsproblem. *Proc. London Math. Soc.* (1937). https://doi.org/10.1112/plms/s2-42.1.230.
21. Church, A. An unsolvable problem of elementary number theory. *Am. J. Math.* (1936). https://doi.org/10.2307/2371045.
22. Post, E. L. Finite combinatory processes—formulation 1. *J. Symb. Log.* **1**, 103–105 (1936).
23. Pour-El, M. B. & Richards, J. I. *Computability in Analysis and Physics: Perspectives in Logic* (Cambridge University Press, 2017). https://doi.org/10.1017/9781316717325.
24. Pour-El, M. B. & Richards, I. The wave equation with computable initial data such that its unique solution is not computable. *Adv. Math. (N. Y.)* **39**, 215–239 (1981).
25. Penrose, R. Précis of *The Emperor's New Mind: Concerning Computers, Minds, and the Laws of Physics. Behav. Brain Sci.* (1990). https://doi.org/10.1017/s0140525x00080675.
26. Roli, A., Jaeger, J. & Kauffman, S. A. How organisms come to know the world: Fundamental limits on artificial general intelligence. *Front. Ecol. Evol.* **9** (2022).
27. Dauben, J. W. Georg Cantor and Pope Leo XIII: Mathematics, theology, and the infinite. *J. Hist. Ideas* (1977). https://doi.org/10.2307/2708842.
28. Black, D. Beating floating point at its own game: Posit arithmetic. *Inside HPC* (2017). https://insidehpc.com/2017/08/beating-floating-point-game-posit-arithmetic/.
29. Gustafson, J. *The End of Error: Unum Computing* (Chapman & Hall, 2015).
30. Lorenz, E. Predictability: Does the flap of a butterfly's wings in Brazil set off a tornado in Texas? *Am. Assoc. Adv. Sci., 139th Meeting* (paper presented December 29, 1972).
31. Lorenz, E. N. Deterministic nonperiodic flow. *J. Atmos. Sci.* (1963). https://doi.org/10.1175/1520-0469(1963)020<0130:DNF>2.0.CO;2.
32. Coveney, P. V. & Wan, S. On the calculation of equilibrium thermodynamic properties from molecular dynamics. *Phys. Chem. Chem. Phys.* (2016). https://doi.org/10.1039/c6cp02349e.
33. Hardaker, P. Weather in my life—Professor Tim Palmer FRS, president of the RMetS. *Weather* (2011). https://doi.org/10.1002/wea.814.
34. Boghosian, B. M., Fazendeiro, L. M., Lätt, J., Tang, H. & Coveney, P. V. New variational principles for locating periodic orbits of differential equations. *Philos. Trans. R. Soc. A Math. Phys. Eng. Sci.* **369**, 2211–2218 (2011).
35. Smith, J. H. et al. How neurons exploit fractal geometry to optimize their network connectivity. *Sci. Rep.* **11**, 2332 (2021).
36. Cvitanović, P. Periodic orbits as the skeleton of classical and quantum chaos. *Phys. D Nonlinear Phenom.* **51**, 138–151 (1991).
37. Coveney, P. V. & Wan, S. On the calculation of equilibrium thermodynamic properties from molecular dynamics. *Phys. Chem. Chem. Phys.* (2016). https://doi.org/10.1039/c6cp02349e.
38. Nee, S. Survival and weak chaos. *R. Soc. Open Sci.* (2018). https://doi.org/10.1098/rsos.172181.
39. Boghosian, B. M., Coveney, P. V. & Wang, H. A. New pathology in the simulation of chaotic dynamical systems on digital computers. *Adv. Theory Simulations* (2019). https://doi.org/10.1002/adts.201900125.
40. Sauer, T. D. Shadowing breakdown and large errors in dynamical simulations of physical systems. *Phys. Rev. E* **65**, 36220 (2002).
41. Sauer, T. Computer arithmetic and sensitivity of natural measure. *J. Differ. Equations Appl.* **11**, 669–676 (2005).
42. Pool, R. Is it healthy to be chaotic? *Science* (1989) https://doi.org/10.1126/science.2916117.
43. May, R. M. Uses and abuses of mathematics in biology. *Science* (2004). https://doi.org/10.1126/science.1094442.

44. Highfield, R. Ramanujan: Divining the origins of genius. *Science Museum Blog* (2016). https://blog.sciencemuseum.org.uk/ramanujan-divining-the-origins-of-genius/.

45. Herron, M. D. & Doebeli, M. Parallel evolutionary dynamics of adaptive diversification in *Escherichia coli. PLoS Biol.* (2013). https://doi.org/10.1371/journal.pbio.1001490.

46. Wiser, M. J., Ribeck, N. & Lenski, R. E. Long-term dynamics of adaptation in asexual populations. *Science* (2013). https://doi.org/10.1126/science.1243357.

47. Łuksza, M. & Lässig, M. A predictive fitness model for influenza. *Nature* (2014). https://doi.org/10.1038/nature13087.

48. Nowak, M. A. & Highfield, R. *Supercooperators: Altruism, Evolution, and Why We Need Each Other to Succeed* (Canongate, 2011).

49. Russell, B., Slater, J. G. & Frohmann, B. *Logical and Philosophical Papers, 1909–13*, vol. 6 (Routledge, 1992).

50. Weinan, E. The dawning of a new era in applied mathematics. *Not. Am. Math. Soc.* **68**, 565–571 (2021).

Chapter 3: From Analogue to Digital You

1. Feynman, R. P. *The Feynman Lectures on Physics* (Addison-Wesley, 1963–1965).

2. Marchant, J. *The Human Cosmos* (Dutton, 2020).

3. Harlow, F. H. & Metropolis, N. Computing & computers: Weapons simulation leads to the computer era. *Los Alamos Sci.*, 132–141 (Winter/spring 1983).

4. Anderson, H. Metropolis, Monte Carlo and the MANIAC. *Los Alamos Sci.*, 96–107 (February 1986).

5. Schweber, S. S. *In the Shadow of the Bomb: Oppenheimer, Bethe, and the Moral Responsibility of the Scientist* (Princeton University Press, 2013).

6. Rhodes, R. *The Making of the Atomic Bomb* (Simon & Schuster, 1986).

7. Metropolis, N., Rosenbluth, A. W., Rosenbluth, M. N., Teller, A. H. & Teller, E. Equation of state calculations by fast computing machines. *J. Chem. Phys.* **21**, 1087–1092 (1953).

8. Battimelli, G. & Ciccotti, G. Berni Alder and the pioneering times of molecular simulation. *Eur. Phys. J. H* **43** (2018).

9. Alder, B. J. & Wainwright, T. E. Phase transition for a hard sphere system. *J. Chem. Phys.* **27**, 1208–1209 (1957).

10. Moore, J. W. A personal view of the early development of computational neuroscience in the USA. *Front. Comput. Neurosci.* (2010). https://doi.org/10.3389/fncom.2010.00020.

11. Huxley, A. F. Ion movements during nerve activity. *Ann. N. Y. Acad. Sci.* **81**, 221–246 (1959).

12. Chen, Z. & Auffray, C. (eds.). *The Selected Papers of Denis Noble CBE FRS: A Journey in Physiology towards Enlightenment* (Imperial College Press, 2012).

13. Noble, D. A modification of the Hodgkin-Huxley equations applicable to Purkinje fibre action and pacemaker potentials. *J. Physiol.* (1962). https://doi.org/10.1113/jphysiol.1962.sp006849.

14. McAllister, R. E., Noble, D. & Tsien, R. W. Reconstruction of the electrical activity of cardiac Purkinje fibres. *J. Physiol.* (1975). https://doi.org/10.1113/jphysiol.1975.sp011080.

15. DiFrancesco, D. & Noble, D. A model of cardiac electrical activity incorporating ionic pumps and concentration changes. *Philos. Trans. R. Soc. B Biol. Sci.* (1985). https://doi.org/10.1098/rstb.1985.0001.

16. Beeler, G. W. & Reuter, H. Reconstruction of the action potential of ventricular myocardial fibres. *J. Physiol.* (1977). https://doi.org/10.1113/jphysiol.1977.sp011853.

17. Noble, D. Successes and failures in modeling heart cell electrophysiology. *Heart Rhythm* (2011). https://doi.org/10.1016/j.hrthm.2011.06.014.

18. Tomayko, J. Computers in spaceflight: The NASA experience. In *Encyclopedia of Computer Science and Technology* (eds. Kent, A. & Williams, J. G.), vol. 18 (CRC Press, 1987). https://history.nasa.gov/computers/Ch2-5.html.

19. Hey, T. & Papay, G. *The Computing Universe: A Journey through a Revolution* (Cambridge University Press, 2014).

20. Kaufmann, W. J. & Smarr, L. L. Supercomputing and the transformation of science. *La Météorologie* (1995). https://doi.org/10.4267/2042/52011.

21. Murray, C. *The Supermen: The Story of Seymour Cray and the Technical Wizards behind the Supercomputer* (Wiley, 1997).

22. Moore, G. E. Cramming more components onto integrated circuits. *IEEE Solid-State Circuits Soc. Newsl.* (2009). Reprinted from *Electronics* **38** (8): 114ff. (April 19, 1965). https://doi.org/10.1109/n-ssc.2006.4785860.

23. Dennard, R. H. et al. Design of ion-implanted MOSFET's with very small physical dimensions. *IEEE J. Solid-State Circuits* (1974). https://doi.org/10.1109/JSSC.1974.1050511.

24. Alowayyed, S., Groen, D., Coveney, P. V. & Hoekstra, A. G. Multiscale computing in the exascale era. *J. Comput. Sci.* (2017). https://doi.org/10.1016/j.jocs.2017.07.004.

25. Kogge, P. et al. *ExaScale Computing Study: Technology Challenges in Achieving Exascale Systems*. Def. Adv. Res. Proj. Agency Inf. Process. Tech. Off. (DARPA IPTO), Technical Represent. **15** (2008).

26. Mirhoseini, A. et al. A graph placement methodology for fast chip design. *Nature* **594**, 207–212 (2021).

27. Lee, C. T. & Amaro, R. E. Exascale computing: A new dawn for computational biology. *Comput. Sci. Eng.* (2018). https://doi.org/10.1109/MCSE.2018.05329812.

28. Nievergelt, J. Parallel methods for integrating ordinary differential equations. *Commun. ACM* **7**, 731–733 (1964).

29. Lions, J.-L., Maday, Y. & Turinici, G. Résolution d'EDP par un schéma en temps «pararéel». *Comptes Rendus l'Académie des Sci.—Ser. I—Math.* **332**, 661–668 (2001).

30. Eames, I. & Flor, J. B. New developments in understanding interfacial processes in turbulent flows. *Philos. Trans. R. Soc. A Math. Phys. Eng. Sci.* **369**, 702–705 (2011).

31. Boghosian, B. M. et al. Unstable periodic orbits in the Lorenz attractor. *Philos. Trans. R. Soc. A Math. Phys. Eng. Sci.* **369**, 2345–2353 (2011).

32. Jia, W. et al. Pushing the limit of molecular dynamics with ab initio accuracy to 100 million atoms with machine learning. *Proc. Int. Conf. High Performance Comput., Networking, Storage and Analysis* (IEEE, 2020).

33. Zimmerman, M. I. et al. Citizen scientists create an exascale computer to combat COVID-19. *bioRxiv* (2020). https://doi.org/10.1101/2020.06.27.175430.

34. Mann, A. Nascent exascale supercomputers offer promise, present challenges. *Proc. Natl. Acad. Sci. U. S. A.* (2020). https://doi.org/10.1073/pnas.2015968117.

35. Oak Ridge National Laboratory. Frontier supercomputer debuts as world's fastest, breaking exascale barrier. Press release (May 30, 2022). https://www.ornl.gov/news/frontier-supercomputer-debuts-worlds-fastest-breaking-exascale-barrier.

36. Golaz, J. C. et al. The DOE E3SM coupled model version 1: Overview and evaluation at standard resolution. *J. Adv. Model. Earth Syst.* (2019). https://doi.org/10.1029/2018MS001603.

37. Wagman, B. M., Lundquist, K. A., Tang, Q., Glascoe, L. G. & Bader, D. C. Examining the climate effects of a regional nuclear weapons exchange using a multiscale atmospheric modeling approach. *J. Geophys. Res. Atmos.* (2020). https://doi.org/10.1029/2020JD033056.

38. Reed, P. M. & Hadka, D. Evolving many-objective water management to exploit exascale computing. *Water Resour. Res.* (2014). https://doi.org/10.1002/2014WR015976.

39. Degrave, J. et al. Magnetic control of tokamak plasmas through deep reinforcement learning. *Nature* **602**, 414–419 (2022).

40. Kates-Harbeck, J., Svyatkovskiy, A. & Tang, W. Predicting disruptive instabilities in controlled fusion plasmas through deep learning. *Nature* (2019). https://doi.org/10.1038/s41586-019-1116-4.

41. Coveney, P. V, Groen, D. & Hoekstra, A. G. Reliability and reproducibility in computational science: Implementing validation, verification and uncertainty quantification in silico. *Philos. Trans. R. Soc. A Math. Phys. Eng. Sci.* **379**, 20200409 (2021).

42. Coveney, P. V & Highfield, R. R. When we can trust computers (and when we can't). *Philos. Trans. R. Soc. A Math. Phys. Eng. Sci.* **379**, 20200067 (2021).

43. Johnson, N. F. et al. The online competition between pro- and anti-vaccination views. *Nature* **582**, 230–233 (2020).

44. Perkel, J. M. Challenge to scientists: Does your ten-year-old code still run? *Nature* **584**, 656–658 (2020).

45. VECMA. https://www.vecma.eu/.

46. VECMA Toolkit. https://www.vecma-toolkit.eu/.

47. Highfield, R. R. Coronavirus: Virtual pandemics. *Science Museum Group Blog* (2020). https://www.sciencemuseumgroup.org.uk/blog/coronavirus-virtual-pandemics/.

48. Edeling, W. et al. The impact of uncertainty on predictions of the CovidSim epidemiological code. *Nat. Comput. Sci.* **1**, 128–135 (2021).

49. Jordan, J. et al. Extremely scalable spiking neuronal network simulation code: From laptops to exascale computers. *Front. Neuroinform.* (2018). https://doi.org/10.3389/fninf.2018.00002.

50. Hernandez-Boussard, T. et al. Digital twins for predictive oncology will be a paradigm shift for precision cancer care. *Nat. Med.* **27**, 2065–2066 (2021).

51. Bhattacharya, T. et al. AI meets exascale computing: Advancing cancer research with large-scale high performance computing. *Front. Oncol.* **9** (2019).

Chapter 4: Big AI

1. IBM. *What Will We Make of This Moment?* Annual report (2013). https://www.ibm.com/annualreport/2013/bin/assets/2013_ibm_annual.pdf.

2. IBM. 10 key marketing trends for 2017. IBM.com (December 2016). https://app.box.com/s/ez6qv90o6o2txk1fq69spc03b2l3iehd.

3. Anderson, C. The end of theory: The data deluge makes the scientific method obsolete. *Wired Mag.* (2008). https://doi.org/10.1016/j.ecolmodel.2009.09.008.

4. Samuel, A. L. Some studies in machine learning. *IBM J. Res. Dev.* (1959).

5. McCulloch, W. S. & Pitts, W. A logical calculus of the ideas immanent in nervous activity. *Bull. Math. Biophys.* (1943). https://doi.org/10.1007/BF02478259.

6. Hutson, M. Robo-writers: The rise and risks of language-generating AI. *Nature* **591**, 22–25 (2021).

7. Wurman, P. R. et al. Outracing champion Gran Turismo drivers with deep reinforcement learning. *Nature* **602**, 223–228 (2022).

8. Heider, F. & Simmel, M. An experimental study of apparent behavior. *Am. J. Psychol.* **57**, 243–259 (1944).

9. Silver, D. et al. A general reinforcement learning algorithm that masters chess, shogi, and Go through self-play. *Science* (2018) https://doi.org/10.1126/science.aar6404.

10. Silver, D. et al. Mastering the game of Go without human knowledge. *Nature* (2017). https://doi.org/10.1038/nature24270.

11. Goodfellow, I. J. et al. Generative adversarial nets. *Adv. Neural Inf. Process. Sys.* (2014). https://proceedings.neurips.cc/paper/2014/file/5ca3e9b122f61f8f06494c97b1afccf3-Paper.pdf.

12. Zhang, H. et al. StackGAN++: Realistic image synthesis with stacked generative adversarial networks. *IEEE Trans. Pattern Anal. Mach. Intell.* (2019). https://doi.org/10.1109/TPAMI .2018.2856256.

13. Davies, A. et al. Advancing mathematics by guiding human intuition with AI. *Nature* **600**, 70–74 (2021); Fawzi, A., Balog, M., Huang, A. et al. Discovering faster matrix multiplication algorithms with reinforcement learning. *Nature* **610**, 47–53 (2022). https://doi.org/10.1038/s41586-022-05172-4.

14. Weinan E. The dawning of a new era in applied mathematics. *Notices Am. Math. Soc.*, **68**(4), 565–571 (2021).

15. Cheng, B., Engel, E. A., Behler, J., Dellago, C. & Ceriotti, M. Ab initio thermodynamics of liquid and solid water. *Proc. Natl. Acad. Sci.* **116**, 1110–1115 (2019).

16. Lehman, C. D. et al. Diagnostic accuracy of digital screening mammography with and without computer-aided detection. *JAMA Intern. Med.* (2015). https://doi.org/10.1001 /jamainternmed.2015.5231.

17. McKinney, S. M. et al. International evaluation of an AI system for breast cancer screening. *Nature* (2020). https://doi.org/10.1038/s41586-019-1799-6.

18. Lu, M. Y. et al. AI-based pathology predicts origins for cancers of unknown primary. *Nature* **594**, 106–110 (2021).

19. Roberts, M. et al. Common pitfalls and recommendations for using machine learning to detect and prognosticate for COVID-19 using chest radiographs and CT scans. *Nat. Mach. Intell.* **3**, 199–217 (2021).

20. Mei, X. et al. Artificial intelligence–enabled rapid diagnosis of patients with COVID-19. *Nat. Med.* **26**, 1224–1228 (2020).

21. Ng, A. X-rays: The AI hype. *IEEE Spectrum* (2021). https://spectrum.ieee.org/andrew -ng-xrays-the-ai-hype.

22. Müller, U., Ivlev, S., Schulz, S. & Wölper, C. Automated crystal structure determination has its pitfalls: Correction to the crystal structures of iodine azide. *Angew. Chemie Int. Ed.* (2021). https://doi.org/10.1002/anie.202105666.

23. Jumper, J. et al. Highly accurate protein structure prediction with AlphaFold. *Nature* (2021). https://doi.org/10.1038/S41586-021-03819-2.

24. Callaway, E. 'The entire protein universe': AI predicts shape of nearly every known protein. *Nature* **608**, 15–16 (2022).

25. Baek, M. et al. Accurate prediction of protein structures and interactions using a three-track neural network. *Science* (2021) https://doi.org/10.1126/science.abj8754.

26. Workman, P. The drug discoverer—reflecting on DeepMind's AlphaFold artificial intelligence success—what's the real significance for protein folding research and drug discovery? Institute of Cancer Research, London (2021). https://www.icr.ac.uk/blogs/ the-drug-discoverer/page-details/reflecting-on-deepmind-s-alphafold-artificial-intelligence-success-what-s-the-real-significance-for-protein-folding-research-and-drug -discovery.

27. Szegedy, C. et al. Intriguing properties of neural networks. *2nd Int. Conf. Learning Representations—Conf. Track Proc.* (ICLR, 2014).

28. YouTube. A DARPA perspective on artificial intelligence (posted February 15, 2017). https://www.youtube.com/watch?v=-O01G3tSYpU.

29. Eykholt, K. et al. Robust physical-world attacks on deep learning visual classification. *Proc. IEEE Comput. Soc. Conf.Comput. Vision and Pattern Recognition* (2018). https://doi.org/10.1109/CVPR.2018.00175.

30. Highfield, R. Bill Gates and will.i.am argue for progress through investment in science. *Science Museum Blog* (2016). https://blog.sciencemuseum.org.uk/bill-gates-and-will-i-am-argue-for-progress-through-investment-in-science/.

31. Hawkins, D. M. The problem of overfitting. *J. Chem. Inf. Comput. Sci.* (2004). https://doi.org/10.1021/ci0342472.

32. Coveney, P. V. & Highfield, R. R. From digital hype to analogue reality: Universal simulation beyond the quantum and exascale eras. *J. Comput. Sci.* (2020). https://doi.org/10.1016/j.jocs.2020.101093.

33. Pathak, J., Lu, Z., Hunt, B. R., Girvan, M. & Ott, E. Using machine learning to replicate chaotic attractors and calculate Lyapunov exponents from data. *Chaos* (2017). https://doi.org/10.1063/1.5010300.

34. Pathak, J., Hunt, B., Girvan, M., Lu, Z. & Ott, E. Model-free prediction of large spatio-temporally chaotic systems from data: A reservoir computing approach. *Phys. Rev. Lett.* (2018). https://doi.org/10.1103/PhysRevLett.120.024102.

35. Federrath, C., Klessen, R. S., Iapichino, L. & Beattie, J. R. The sonic scale of interstellar turbulence. *Nat. Astron.* (2021). https://doi.org/10.1038/s41550-020-01282-z.

36. Wagner, G. & Weitzman, M. *Climate Shock* (Princeton University Press, 2016).

37. Taleb, N. N. The black swan: Why don't we learn that we don't learn? Paper presented at the US Department of Defense Highland Forum, Las Vegas (2004).

38. Succi, S. & Coveney, P. V. Big data: The end of the scientific method? *Philos. Trans. R. Soc. A Math. Phys. Eng. Sci.* (2019). https://doi.org/10.1098/rsta.2018.0145.

39. Wan, S., Sinclair, R. C. & Coveney, P. V. Uncertainty quantification in classical molecular dynamics. *Philos. Trans. R. Soc. A Math. Phys. Eng. Sci.* **379**, 20200082 (2021).

40. Lakshminarayanan, B., Pritzel, A. & Blundell, C. Simple and scalable predictive uncertainty estimation using deep ensembles. *arXiv* 1612.01474 (2016).

41. Hinton, G. Boltzmann machines. *Encyclopedia of Machine Learning and Data Mining* (2017). https://doi.org/10.1007/978-1-4899-7687-1_31.

42. García-Martín, E., Rodrigues, C. F., Riley, G. & Grahn, H. Estimation of energy consumption in machine learning. *J. Parallel Distrib. Comput.* **134**, 75–88 (2019).

43. Thompson, N. C., Greenewald, K. & Lee, K. Deep learning's diminishing returns. *IEEE Spectrum* (2021). https://spectrum.ieee.org/deep-learning-computational-cost.

44. Wright, L. G. et al. Deep physical neural networks trained with backpropagation. *Nature* **601**, 549–555 (2022).

45. Succi, S. & Coveney, P. V. Big data: The end of the scientific method? *Philos. Trans. R. Soc. A Math. Phys. Eng. Sci.* (2019). https://doi.org/10.1098/rsta.2018.0145.

46. Calude, C. S. & Longo, G. The deluge of spurious correlations in big data. *Found. Sci.* (2017). https://doi.org/10.1007/s10699-016-9489-4.

47. Zernicka-Goetz, M. & Highfield, R. *The Dance of Life* (W. H. Allen, 2020).

48. Choudhary, A., Fox, G. & Hey, T. (eds.). *AI for Science* (World Scientific, in press 2022).

49. Karniadakis, G. E. et al. Physics-informed machine learning. *Nat. Rev. Phys.* **3**, 422–440 (2021).

50. Erge, O. & van Oort, E. Combining physics-based and data-driven modeling in well construction: Hybrid fluid dynamics modeling. *J. Nat. Gas Sci. Eng.* **97**, 104348 (2022).

51. Davis, J. J. et al. Antimicrobial resistance prediction in PATRIC and RAST. *Sci. Rep.* **6**, 27930 (2016).

52. McSkimming, D. I., Rasheed, K. & Kannan, N. Classifying kinase conformations using a machine learning approach. *BMC Bioinformatics* **18**, 86 (2017).

53. Rufa, D. A. et al. Towards chemical accuracy for alchemical free energy calculations with hybrid physics-based machine learning/molecular mechanics potentials. *bioRxiv* (2020). https://doi.org/10.1101/2020.07.29.227959.

54. Bhati, A. P. et al. Pandemic drugs at pandemic speed: Infrastructure for accelerating COVID-19 drug discovery with hybrid machine learning and physics-based simulations on high-performance computers. *Interface Focus* **11**, 20210018 (2021).

55. Wan, S., Bhati, A. P., Wade, A. D., Alfè, D. & Coveney, P. V. Thermodynamic and structural insights into the repurposing of drugs that bind to SARS-CoV-2 main protease. *Mol. Syst. Des. Eng.* (2022). https://doi.org/10.1039/D1ME00124H.

56. Clyde, A. et al. High-throughput virtual screening and validation of a SARS-CoV-2 main protease noncovalent inhibitor. *J. Chem. Inf. Modeling* **62** (1), 116–128 (2022).

57. Raissi, M., Perdikaris, P. & Karniadakis, G. E. Physics-informed neural networks: A deep learning framework for solving forward and inverse problems involving nonlinear partial differential equations. *J. Comput. Phys.* **378**, 686–707 (2019).

58. Kharazmi, E. et al. Identifiability and predictability of integer- and fractional-order epidemiological models using physics-informed neural networks. *Nat. Comput. Sci.* **1**, 744–753 (2021).

59. Alber, M., et al. Integrating machine learning and multiscale modeling—perspectives, challenges, and opportunities in the biological, biomedical, and behavioral sciences. *Npj Digit. Med.* **2**, 115 (2019).

Chapter 5: A Simulating Life

1. Einstein, A. *The Ultimate Quotable Einstein* (Princeton University Press, 2016).

2. Highfield, R. & Carter, P. *The Private Lives of Albert Einstein* (Faber, 1993).

3. Wheeler, J. A. *A Journey into Gravity and Spacetime* (Scientific American Library/W. H. Freeman, 1990).

4. Clarke, B. Normal bone anatomy and physiology. *Clin. J. Amer. Soc. Nephrol.* (2008). https://doi.org/10.2215/CJN.04151206.

5. Brooks, S. V. Current topics for teaching skeletal muscle physiology. *Amer. J. Physiol.– Adv. Physiol. Edu.* (2003). https://doi.org/10.1152/advan.2003.27.4.171.

6. Chen, Z. & Auffray, C. (eds.). *The Selected Papers of Denis Noble CBE FRS: A Journey in Physiology towards Enlightenment* (Imperial College Press, 2012).

7. Wei, F. et al. Stress fiber anisotropy contributes to force-mode dependent chromatin stretching and gene upregulation in living cells. *Nat. Commun.* **11**, 4902 (2020).

8. Noble, D. *The Music of Life: Biology beyond the Genome* (Oxford University Press, 2006).

9. Noble, D. Claude Bernard, the first systems biologist, and the future of physiology. *Exp. Physiol.* (2008). https://doi.org/10.1113/expphysiol.2007.038695.

10. Novikoff, A. B. The concept of integrative levels and biology. *Science* (1945). https://doi.org/10.1126/science.101.2618.209.

11. Srinivasan, B. A guide to the Michaelis-Menten equation: Steady state and beyond. *FEBS J.* (2021). https://doi.org/10.1111/febs.16124.

12. Anderson, H. Metropolis, Monte Carlo and the MANIAC. *Los Alamos Sci.* 96–107 (1986). https://permalink.lanl.gov/object/tr?what=info:lanl-repo/lareport/LA-UR-86-2600-05.

13. Clark, A. J. Post on *DC's Improbable Science* website (2008). http://www.dcscience.net /tag/ajclark/.
14. Clark, A. J. The reaction between acetyl choline and muscle cells. *J. Physiol.* **61**, 530–546 (1926).
15. Dance, A. Beyond coronavirus: The virus discoveries transforming biology. *Nature* **595**, 22–25 (2021).
16. Gibb, R. et al. Zoonotic host diversity increases in human-dominated ecosystems. *Nature* (2020). https://doi.org/10.1038/s41586-020-2562-8.
17. Nakane, T. et al. Single-particle cryo-EM at atomic resolution. *Nature* **587**, 152–156 (2020).
18. Singer, A. & Sigworth, F. J. Computational methods for single-particle cryo-EM. *Ann. Rev.Biomed. Data Sci.* **3**(1), 163–190 (2020).
19. Sharp, P. M. & Hahn, B. H. Origins of HIV and the AIDS pandemic. *Cold Spring Harb. Perspect. Med.* (2011). https://doi.org/10.1101/cshperspect.a006841.
20. Korber, B. et al. Timing the ancestor of the HIV-1 pandemic strains. *Science* (2000). https://doi.org/10.1126/science.288.5472.1789.
21. Karplus, M. & McCammon, J. A. Molecular dynamics simulations of biomolecules. *Nature Structural Biology* (2002). https://doi.org/10.1038/nsb0902–646.
22. Stein, M., Gabdoulline, R. R. & Wade, R. C. Bridging from molecular simulation to biochemical networks. *Curr. Opinion in Structural Biol.* (2007). https://doi.org/10.1016/j .sbi.2007.03.014.
23. Smock, R. G. & Gierasch, L. M. Sending signals dynamically. *Science* (2009). https://doi .org/10.1126/science.1169377.
24. Shaw, D. E. et al. Anton 2: Raising the bar for performance and programmability in a special-purpose molecular dynamics supercomputer. *SC '14: Proc. Int. Conf. High Performance Comput., Networking, Storage and Analysis*, 41–53 (2014). https://doi.org/10.1109 /SC.2014.9.
25. Di Natale, F. et al. A massively parallel infrastructure for adaptive multiscale simulations: Modeling RAS initiation pathway for cancer. *Proc. Int. Conf. High Performance Comput., Networking, Storage and Analysis* (ACM, 2019). https://doi.org/10.1145/3295500 .3356197.
26. Könnyu, B. et al. Gag-Pol processing during HIV-1 virion maturation: A systems biology approach. *PLoS Comput. Biol.* (2013). https://doi.org/10.1371/journal.pcbi.1003103.
27. Reddy, B. & Yin, J. Quantitative intracellular kinetics of HIV type 1. *AIDS Res. Hum. Retroviruses* (1999). https://doi.org/10.1089/088922299311457.
28. Charlotte Eccleston, R., Wan, S., Dalchau, N. & Coveney, P. V. The role of multiscale protein dynamics in antigen presentation and T lymphocyte recognition. *Front. Immunol.* (2017). https://doi.org/10.3389/fimmu.2017.00797.
29. Kim, H. & Yin, J. Effects of RNA splicing and post-transcriptional regulation on HIV-1 growth: A quantitative and integrated perspective. *IEEE Proc. Sys. Biol.* (2005). https:// doi.org/10.1049/ip-syb:20050004.
30. Wang, Y. & Lai, L. H. Modeling the intracellular dynamics for Vif-APO mediated HIV-1 virus infection. *Chinese Sci. Bull.* (2010). https://doi.org/10.1007/s11434-010-3103-x.
31. Dalchau, N. et al. A peptide filtering relation quantifies MHC class I peptide optimization. *PLoS Comput. Biol.* (2011). https://doi.org/10.1371/journal.pcbi.1002144.
32. Calis, J.J.A. et al. Properties of MHC class I presented peptides that enhance immunogenicity. *PLoS Comput. Biol.* (2013). https://doi.org/10.1371/journal.pcbi.1003266.
33. Lever, M. et al. Architecture of a minimal signaling pathway explains the T-cell response to a 1 million-fold variation in antigen affinity and dose. *Proc. Natl. Acad. Sci. U. S. A.* (2016). https://doi.org/10.1073/pnas.1608820113.

34. Eccleston, R. C., Coveney, P. V. & Dalchau, N. Host genotype and time dependent antigen presentation of viral peptides: Predictions from theory. *Sci. Rep.* (2017). https://doi.org /10.1038/s41598-017-14415-8.

35. Boulanger, D.S.M. et al. A mechanistic model for predicting cell surface presentation of competing peptides by MHC class I molecules. *Front. Immunol.* **9**, 1538 (2018).

36. Wan, S., Flower, D. R. & Coveney, P. V. Toward an atomistic understanding of the immune synapse: Large-scale molecular dynamics simulation of a membrane-embedded TCR-pMHC-CD4 complex. *Mol. Immunol.* **45**, 1221–1230 (2008).

37. Eccleston, R. C., Wan, S., Dalchau, N. & Coveney, P. V. The role of multiscale protein dynamics in antigen presentation and T lymphocyte recognition. *Front. Immunol.* **8**, 797 (2017).

38. Lever, M., Maini, P. K., van der Merwe, P. A. & Dushek, O. Phenotypic models of T cell activation. *Nat. Rev. Immunol.* **14**, 619–629 (2014).

39. Lever, M. et al. Architecture of a minimal signaling pathway explains the T-cell response to a 1 million-fold variation in antigen affinity and dose. *Proc. Natl. Acad. Sci. U. S. A.* (2016). https://doi.org/10.1073/pnas.1608820113.

40. Pettmann, J. et al. The discriminatory power of the T cell receptor. *Elife* **10**, e67092 (2021).

41. Coulson, C. A. Samuel Francis Boys, 1911–1972. *Biogr. Mem. Fellows R. Soc.* **19**, 95–115 (1973).

42. Bullard, E. *Computers and Their Role in the Physical Sciences* (Gordon and Breach, 1970).

43. Christov, C. Z. et al. Conformational effects on the pro-S hydrogen abstraction reaction in cyclooxygenase-1: An integrated QM/MM and MD study. *Biophys. J.* **104**, L5–L7 (2013).

44. Bhati, A. P. & Coveney, P. V. Large scale study of ligand-protein relative binding free energy calculations: Actionable predictions from statistically robust protocols. *ChemRxiv* (2021). https://doi.org/10.26434/chemrxiv-2021-zdzng.

45. DiMasi, J. A., Grabowski, H. G. & Hansen, R. W. Innovation in the pharmaceutical industry: New estimates of R&D costs. *J. Health Econ.* (2016). https://doi.org/10.1016/j .jhealeco.2016.01.012.

46. Turner, E. H., Matthews, A. M., Linardatos, E., Tell, R. A. & Rosenthal, R. Selective publication of antidepressant trials and its influence on apparent efficacy. *N. Engl. J. Med.* (2008). https://doi.org/10.1056/NEJMsa065779.

47. Jefferson, T. et al. Neuraminidase inhibitors for preventing and treating influenza in adults and children. *Cochrane Database Sys. Rev.* (2014). https://doi.org/10.1002 /14651858.CD008965.pub4.

48. Leucht, S., Helfer, B., Gartlehner, G. & Davis, J. M. How effective are common medications?: A perspective based on meta-analyses of major drugs. *BMC Med.* (2015). https:// doi.org/10.1186/s12916-015-0494-1.

49. Bohacek, R. S., McMartin, C. & Guida, W. C. The art and practice of structure-based drug design: A molecular modeling perspective. *Med. Res. Rev.* **16**, 3–50 (1996).

50. Wright, D. W., Hall, B. A., Kenway, O. A., Jha, S. & Coveney, P. V. Computing clinically relevant binding free energies of HIV-1 protease inhibitors. *J. Chem. Theory Comput.* (2014). https://doi.org/10.1021/ct4007037.

51. Vassaux, M., Wan, S., Edeling, W. & Coveney, P. V. Ensembles are required to handle aleatoric and parametric uncertainty in molecular dynamics simulation. *J. Chem. Theory Comput.* (2021). https://doi.org/10.1021/acs.jctc.1c00526.

52. Wan, S. et al. Rapid and reliable binding affinity prediction of bromodomain inhibitors: A computational study. *J. Chem. Theory Comput.* (2017). https://doi.org/10.1021/acs.jctc .6b00794.

53. Wan, S., Bhati, A. P., Zasada, S. J. & Coveney, P. V. Rapid, accurate, precise and reproducible ligand–protein binding free energy prediction. *Interface Focus* **10**, 20200007 (2020).

54. Wright, D. W., Hall, B. A., Kenway, O. A., Jha, S. & Coveney, P. V. Computing clinically relevant binding free energies of HIV-1 protease inhibitors. *J. Chem. Theory Comput.* (2014). https://doi.org/10.1021/ct4007037.

55. Sloot, P.M.A. et al. HIV decision support: From molecule to man. *Philos. Trans. R. Soc. A Math. Phys. Eng. Sci.* **367**, 2691–2703 (2009).

56. Fowler, P. W. et al. Robust prediction of resistance to trimethoprim in *Staphylococcus aureus*. *Cell Chem. Biol.* **25**, 339–349 (2018).

57. Fowler, P. W. How quickly can we predict trimethoprim resistance using alchemical free energy methods? *Interface Focus* **10**, 20190141 (2020).

58. Global Pathogen Analysis System. https://gpas.cloud/.

59. Zernicka-Goetz, M. & Highfield, R. *The Dance of Life* (W. H. Allen, 2020).

60. Nowak, M. A. & Highfield, R. *Supercooperators: Altruism, Evolution, and Why We Need Each Other to Succeed* (Canongate, 2011).

61. Bhati, A. P., Wan, S. & Coveney, P. V. Ensemble-based replica exchange alchemical free energy methods: The effect of protein mutations on inhibitor binding. *J. Chem. Theory Comput.* (2019). https://doi.org/10.1021/acs.jctc.8b01118.

62. Wan, S. et al. The effect of protein mutations on drug binding suggests ensuing personalised drug selection. *Sci. Rep.* **11**, 13452 (2021).

63. Sadiq, S. K. et al. Patient-specific simulation as a basis for clinical decision-making. *Philos. Trans. R. Soc. A Math. Phys. Eng. Sci.* (2008). https://doi.org/10.1098/rsta.2008.0100.

64. Passini, E. et al. Human in silico drug trials demonstrate higher accuracy than animal models in predicting clinical pro-arrhythmic cardiotoxicity. *Front. Physiol.* (2017). https://doi.org/10.3389/fphys.2017.00668.

65. Horby, P. et al. Effect of hydroxychloroquine in hospitalized patients with COVID-19: Preliminary results from a multi-centre, randomized, controlled trial. *medRxiv* (2020). https://doi.org/10.1101/2020.07.15.20151852.

66. Aguado-Sierra, J. et al. In-silico clinical trial using high performance computational modeling of a virtual human cardiac population to assess drug-induced arrhythmic risk. *medRxiv* (2021). https://doi.org/10.1101/2021.04.21.21255870.

67. Yang, P.-C. et al. A computational pipeline to predict cardiotoxicity. *Circ. Res.* (2020). https://doi.org/10.1161/circresaha.119.316404.

Chapter 6: The Virtual Cell

1. Hardy, G. H. & Snow, C. P. *A Mathematician's Apology* (Cambridge University Press, 2012).

2. Leduc, S. The mechanism of life. *Arch. Roentgen Ray* (1911). https://doi.org/10.1259/arr.1911.0008.

3. Thompson, D. W. *On Growth and Form* (Cambridge University Press, 1992).

4. Turing, A. M. The chemical basis of morphogenesis. *Philos. Trans. R. Soc. B Biol. Sci.* (1952). https://doi.org/10.1098/rstb.1952.0012.

5. Murray, J. D. *Mathematical Biology: I. An Introduction*, vol. 17 (Springer, 2002).

6. Highfield, R. & Rooney, D. The spirit of Alan Turing. *Science Museum Blog* (2012). https://blog.sciencemuseum.org.uk/the-spirit-of-alan-turing/.

7. Castets, V., Dulos, E., Boissonade, J. & De Kepper, P. Experimental evidence of a sustained standing Turing-type nonequilibrium chemical pattern. *Phys. Rev. Lett.* **64**, 2953 (1990).

8. Liu, R. T., Liaw, S. S. & Maini, P. K. Two-stage Turing model for generating pigment patterns on the leopard and the jaguar. *Phys. Rev. E—Stat. Nonlinear, Soft Matter Phys.* (2006). https://doi.org/10.1103/PhysRevE.74.011914.

9. Hamada, H. In search of Turing in vivo: Understanding nodal and lefty behavior. *Developmental Cell* (2012). https://doi.org/10.1016/j.devcel.2012.05.003.

10. Sick, S., Reinker, S., Timmer, J. & Schlake, T. WNT and DKK determine hair follicle spacing through a reaction-diffusion mechanism. *Science* (2006). https://doi.org/10.1126/science.1130088.

11. Watson, J. D. & Crick, F.H.C. Molecular structure of nucleic acids: A structure for deoxyribose nucleic acid. *Nature* **171**, 737–738 (1953).

12. Babu, K. & Koushika, S. P. Sydney Brenner (1927–2019). *Curr. Sci.* **116**, 2106–2109 (2019).

13. Crick, F.H.C. Project K: "The complete solution of *E. coli.*" *Perspect. Biol. Med.* **17**, 67–70 (1973).

14. Morowitz, H. J. The completeness of molecular biology. *Isr. J. Med. Sci.* **20**, 750–753 (1984).

15. Domach, M. M. & Shuler, M. L. A finite representation model for an asynchronous culture of *E. coli. Biotechnol. Bioeng.* **26**, 877–884 (1984).

16 Shimobayashi, S. F., Ronceray, P., Sanders, D. W., Haataja, M. P. & Brangwynne, C. P. Nucleation landscape of biomolecular condensates. *Nature* (2021). https://doi.org/10.1038/s41586-021-03905-5.

17. Venter, J. C. *A Life Decoded: My Genome, My Life* (Viking, 2007).

18. Tomita, M. et al. E-CELL: Software environment for whole-cell simulation. *Bioinformatics* (1999). https://doi.org/10.1093/bioinformatics/15.1.72.

19. Fraser, C. M. et al. The minimal gene complement of Mycoplasma genitalium. *Science* **270**, 397–403 (1995).

20. Karr, J. R. et al. A whole-cell computational model predicts phenotype from genotype. *Cell* (2012). https://doi.org/10.1016/j.cell.2012.05.044.

21. Glass, J. I. et al. Essential genes of a minimal bacterium. *Proc. Natl. Acad. Sci. U. S. A.* (2006). https://doi.org/10.1073/pnas.0510013103.

22. Sanghvi, J. C. et al. Accelerated discovery via a whole-cell model. *Nat. Methods* **10**, 1192–1195 (2013).

23. Hutchison, C. A. et al. Design and synthesis of a minimal bacterial genome. *Science* (2016). https://doi.org/10.1126/science.aad6253.

24. Highfield, R. J. Craig Venter sequenced the human genome. Now he wants to convert DNA into a digital signal. *Wired UK* (2013). https://www.wired.co.uk/article/j-craig-venter-interview.

25. Thornburg, Z. R. et al. Fundamental behaviors emerge from simulations of a living minimal cell. *Cell* **185**, 345–360 (2022).

26. Macklin, D. N. et al. Simultaneous cross-evaluation of heterogeneous *E. coli* datasets via mechanistic simulation. *Science* (2020). https://doi.org/10.1126/science.aav3751.

27. Bachmair, A., Finley, D. & Varshavsky, A. In vivo half-life of a protein is a function of its amino-terminal residue. *Science* **234**, 179–186 (1986).

28. Johnson, G. T. et al. CellPACK: A virtual mesoscope to model and visualize structural systems biology. *Nat. Methods* **12**, 85–91 (2015).

29. Wodke, J.A.H. et al. MyMpn: A database for the systems biology model organism Mycoplasma pneumoniae. *Nucleic Acids Res.* **43**, D618–D623 (2015).

30. Schaff, J., Fink, C. C., Slepchenko, B., Carson, J. H. & Loew, L. M. A general computational framework for modeling cellular structure and function. *Biophys. J.* **73**, 1135–1146 (1997).

31. Neves, S. R. Developing models in Virtual Cell. *Sci. Signal.* (2011). https://doi.org/10.1126/scisignal.2001970.

32. Falkenberg, C. V. et al. Fragility of foot process morphology in kidney podocytes arises from chaotic spatial propagation of cytoskeletal instability. *PloS Comput. Biol.* **13**, e1005433 (2017).

33. Singla, J. et al. Opportunities and challenges in building a spatiotemporal multi-scale model of the human pancreatic β cell. *Cell* (2018). https://doi.org/10.1016/j.cell.2018.03.014.

34. Novak, I. L. & Slepchenko, B. M. A conservative algorithm for parabolic problems in domains with moving boundaries. *J. Comput. Phys.* **270**, 203–213 (2014).

35. Cowan, A. E., Mendes, P. & Blinov, M. L. ModelBricks—modules for reproducible modeling improving model annotation and provenance. *Npj Syst. Biol. Appl.* (2019). https://doi.org/10.1038/s41540-019-0114-3.

36. Nurse, P. Systems biology: Understanding cells. *Nature* **424**, 883 (2003).

37. Klumpe, H. et al. The context-dependent, combinatorial logic of BMP signaling. *bioRxiv* (2020). https://doi.org/10.1101/2020.12.08.416503.

38. Nurse, P. & Hayles, J. The cell in an era of systems biology. *Cell* **144**, 850–854 (2011).

39. Nurse, P. Life, logic and information. *Nature* (2008). https://doi.org/10.1038/454424a.

40. Butters, T. D., et al. Mechanistic links between Na+ channel (SCN5A) mutations and impaired cardiac pacemaking in sick sinus syndrome. *Circ. Res.* **107**, 126–137 (2010).

41. Selvaggio, G. et al. Hybrid epithelial-mesenchymal phenotypes are controlled by microenvironmental factors. *Cancer Res.* (2020). https://doi.org/10.1158/0008-5472.CAN-19-3147.

42. Celada, F. & Seiden, P. E. A computer model of cellular interactions in the immune system. *Immunology Today* (1992). https://doi.org/10.1016/0167-5699(92)90135-T.

43. Shilts, J., Severin, Y., Galaway, F. et al. A physical wiring diagram for the human immune system. *Nature* **608**, 397–404 (2022). https://doi.org/10.1038/s41586-022-05028-x.

44. Celli, S. et al. How many dendritic cells are required to initiate a T-cell response? *Blood* (2012). https://doi.org/10.1182/blood-2012-01-408260.

45. Cockrell, R. C. & An, G. Examining the controllability of sepsis using genetic algorithms on an agent-based model of systemic inflammation. *PloS Comput. Biol.* (2018). https://doi.org/10.1371/journal.pcbi.1005876.

46. Ghaffarizadeh, A., Heiland, R., Friedman, S., Mumenthaler, S. & Macklin, P. PhysiCell: An open source physics-based cell simulator for 3-D multicellular systems. *PloS Comput. Biol.* **14**, e1005991 (2018).

47. Saxena, G., Ponce-de-Leon, M., Montagud, A., Vicente Dorca, D. & Valencia, A. BioFVM-X: An MPI+OpenMP 3-D simulator for biological systems. In *Computational Methods in Systems Biology* (eds. Cinquemani, E. & Paulevé, L.), 266–279 (Springer, 2021).

48. Ozik, J. et al. High-throughput cancer hypothesis testing with an integrated PhysiCell-EMEWS workflow. *BMC Bioinformatics* **19**, 483 (2018).

49. Getz, M. et al. Iterative community-driven development of a SARS-CoV-2 tissue simulator. *bioRxiv* (2021). https://doi.org/10.1101/2020.04.02.019075.

50. Fertig, E. J., Jaffee, E. M., Macklin, P., Stearns, V. & Wang, C. Forecasting cancer: From precision to predictive medicine. *Med* **2**, 1004–1010 (2021).

51. Courtemanche, M. & Winfree, A. T. Re-entrant rotating waves in a Beeler–Reuter based model of two-dimensional cardiac electrical activity. *Int. J. Bifurc. Chaos* (1991). https://doi.org/10.1142/s0218127491000336.

52. Priebe, L. & Beuckelmann, D. J. Simulation study of cellular electric properties in heart failure. *Circ. Res.* (1998). https://doi.org/10.1161/01.RES.82.11.1206.

Chapter 7: How to Create a Human Heart

1. Marage, P. *The Solvay Councils and the Birth of Modern Physics* (Birkhäuser Verlag, 1999). https://doi.org/10.1007/978-3-0348-7703-9.
2. Coveney, P. V., Dougherty, E. R. & Highfield, R. R. Big data need big theory too. *Philos. Trans. R. Soc. A Math. Phys. Eng. Sci.* (2016). https://doi.org/10.1098/rsta.2016.0153.
3. Coveney, P. V., Boon, J. P. & Succi, S. Bridging the gaps at the physics-chemistry-biology interface. *Philos. Trans. R. Soc. A Math. Phys. Eng. Sci.* (2016). https://doi.org/10.1098/rsta.2016.0335.
4. Di Natale, F. et al. A massively parallel infrastructure for adaptive multiscale simulations: Modeling RAS initiation pathway for cancer. *Proc. Int. Conf. High Performance Comput., Networking, Storage and Analysis* (ACM, 2019). https://doi.org/10.1145/3295500.3356197.
5. Saadi, A. A., et al. IMPECCABLE: Integrated modeling pipeline for COVID cure by assessing better leads. *50th Int. Conf. Parallel Process.* (ACM, 2021). https://doi.org/10.1145/3472456.3473524.
6. Wan, S., Bhati, A. P., Zasada, S. J. & Coveney, P. V. Rapid, accurate, precise and reproducible ligand–protein binding free energy prediction. *Interface Focus* **10**, 20200007 (2020).
7. Warshel, A. & Karplus, M. Calculation of ground and excited state potential surfaces of conjugated molecules: I. Formulation and parametrization. *J. Am. Chem. Soc.* **94**, 5612–5625 (1972).
8. Warshel, A. & Levitt, M. Theoretical studies of enzymic reactions: Dielectric, electrostatic and steric stabilization of the carbonium ion in the reaction of lysozyme. *J. Mol. Biol.* **103**, 227–249 (1976).
9. Delgado-Buscalioni, R. & Coveney, P. V. Continuum-particle hybrid coupling for mass, momentum, and energy transfers in unsteady fluid flow. *Phys. Rev. E.–Stat. Nonlinear Soft Matter Phys.* **67**, 46704 (2003).
10. Carrel, A. On the permanent life of tissues outside of the organism. *J. Exp. Med.* **15**, 516–528 (1912).
11. Rudolph, F. et al. Deconstructing sarcomeric structure–function relations in titin-BioID knock-in mice. *Nat. Commun.* (2020). https://doi.org/10.1038/s41467-020-16929-8.
12. Zernicka-Goetz, M. & Highfield, R. *The Dance of Life* (W. H. Allen, 2020).
13. Hunter, P. J. & Borg, T. K. Integration from proteins to organs: The Physiome Project. *Nature Rev. Mol. Cell Biol.* (2003). https://doi.org/10.1038/nrm1054.
14. LeGrice, I. J. et al. Laminar structure of the heart: Ventricular myocyte arrangement and connective tissue architecture in the dog. *Am. J. Physiol.–Heart Circ. Physiol.* (1995). https://doi.org/10.1152/ajpheart.1995.269.2.h571.
15. Chen, Z. & Auffray, C. (eds.). *The Selected Papers of Denis Noble CBE FRS: A Journey in Physiology towards Enlightenment* (Imperial College Press, 2012).
16. Hunter, P. J., McNaughton, P. A. & Noble, D. Analytical models of propagation in excitable cells. *Prog. Biophys. Mol. Biol.* **30**, 99–144 (1975).
17. LeGrice, I. J., Hunter, P. J. & Smaill, B. H. Laminar structure of the heart: A mathematical model. *Am. J. Physiol.–Heart Circ. Physiol.* (1997). https://doi.org/10.1152/ajpheart.1997.272.5.h2466.
18. Smith, N. P., Pullan, A. J. & Hunter, P. J. Generation of an anatomically based geometric coronary model. *Ann. Biomed. Eng.* (2000). https://doi.org/10.1114/1.250.
19. Hunter, P. J., Kohl, P. & Noble, D. Integrative models of the heart: Achievements and limitations. *Philos. Trans. R. Soc. A Math. Phys. Eng. Sci.* (2001). https://doi.org/10.1098/rsta.2001.0816.

20. Chen, Z. & Auffray, C. (eds.). *The Selected Papers of Denis Noble CBE FRS: A Journey in Physiology towards Enlightenment* (Imperial College Press, 2012).
21. Southern, J. et al. Multi-scale computational modelling in biology and physiology. *Prog. Biophys. Mol. Biol.* (2008). https://doi.org/10.1016/j.pbiomolbio.2007.07.019.
22. Baillargeon, B., Rebelo, N., Fox, D. D., Taylor, R. L. & Kuhl, E. The Living Heart Project: A robust and integrative simulator for human heart function. *Eur. J. Mech. A/Solids* (2014). https://doi.org/10.1016/j.euromechsol.2014.04.001.
23. Watanabe, H., Sugiura, S. & Hisada, T. The looped heart does not save energy by maintaining the momentum of blood flowing in the ventricle. *Am. J. Physiol. Heart Circ. Physiol.* **294**, H2191–H2196 (2008).
24. Zhao, J. et al. Three-dimensional integrated functional, structural, and computational mapping to define the structural "fingerprints" of heart-specific atrial fibrillation drivers in human heart ex vivo. *J. Am. Heart Assoc.* (2017). https://doi.org/10.1161/JAHA.117.005922.
25. Corral-Acero, J. et al. The "Digital Twin" to enable the vision of precision cardiology. *Eur. Heart J.* (2020). https://doi.org/10.1093/eurheartj/ehaa159.
26. Chen, Z. et al. Biophysical modeling predicts ventricular tachycardia inducibility and circuit morphology: A combined clinical validation and computer modeling approach. *J. Cardiovasc. Electrophysiol.* (2016). https://doi.org/10.1111/jce.12991.
27. Kayvanpour, E. et al. Towards personalized cardiology: Multi-scale modeling of the failing heart. *PloS One* (2015). https://doi.org/10.1371/journal.pone.0134869.
28. Gillette, K. et al. A framework for the generation of digital twins of cardiac electrophysiology from clinical 12-leads ECGs. *Med. Image Anal.*, 102080 (2021). https://doi.org/10.1016/j.media.2021.102080.
29. Ushenin, K., Kalinin, V., Gitinova, S., Sopov, O. & Solovyova, O. Parameter variations in personalized electrophysiological models of human heart ventricles. *PloS One* **16**, e0249062 (2021).
30. El Faquir, N. et al. Patient-specific computer simulation in TAVR with the self-expanding Evolut R valve. *JACC Cardiovasc. Interv.* (2020). https://doi.org/10.1016/j.jcin.2020.04.018.
31. Morrison, T. M., Dreher, M. L., Nagaraja, S., Angelone, L. M. & Kainz, W. The role of computational modeling and simulation in the total product life cycle of peripheral vascular devices. *J. Med. Devices, Trans. ASME* (2017). https://doi.org/10.1115/1.4035866.
32. Patel, M. R. et al. 1-year impact on medical practice and clinical outcomes of FFRCT: The ADVANCE Registry. *JACC Cardiovasc. Imaging* (2020). https://doi.org/10.1016/j.jcmg.2019.03.003.
33. Baillargeon, B., Rebelo, N., Fox, D. D., Taylor, R. L. & Kuhl, E. The Living Heart Project: A robust and integrative simulator for human heart function. *Eur. J. Mech. A/Solids* (2014). https://doi.org/10.1016/j.euromechsol.2014.04.001.
34. Viceconti, M. et al. *In silico* trials: Verification, validation and uncertainty quantification of predictive models used in the regulatory evaluation of biomedical products. *Methods* **185**, 120–127 (2021).
35. Sahli Costabal, F. et al. Multiscale characterization of heart failure. *Acta Biomater* (2019). https://doi.org/10.1016/j.actbio.2018.12.053.
36. Roney, C. H. et al. Variability in pulmonary vein electrophysiology and fibrosis determines arrhythmia susceptibility and dynamics. *PloS Comput. Biol.* (2018). https://doi.org/10.1371/journal.pcbi.1006166.
37. Boyle, P. M. et al. Computationally guided personalized targeted ablation of persistent atrial fibrillation. *Nat. Biomed. Eng.* **3**, 870–879 (2019).

38. Arevalo, H. J. et al. Arrhythmia risk stratification of patients after myocardial infarction using personalized heart models. *Nat. Commun.* (2016). https://doi.org/10.1038/ncomms11437.

39. Hose, D. R. et al. Cardiovascular models for personalised medicine: Where now and where next? *Med. Eng. Phys.* **72**, 38–48 (2019).

40. Kaboudian, A., Cherry, E. M. & Fenton, F. H. Real-time interactive simulations of large-scale systems on personal computers and cell phones: Toward patient-specific heart modeling and other applications. *Sci. Adv.* **5** (2019).

41. Mandel, W., Oulbacha, R., Roy-Beaudry, M., Parent, S. & Kadoury, S. Image-guided tethering spine surgery with outcome prediction using spatio-temporal dynamic networks. *IEEE Trans. Med. Imaging* **40**, 491–502 (2021).

42. Niederer, S. A., Lumens, J. & Trayanova, N. A. Computational models in cardiology. *Nat. Rev. Cardiol.* **16**, 100–111 (2019).

43. Moss, R. et al. Virtual patients and sensitivity analysis of the Guyton model of blood pressure regulation: Towards individualized models of whole-body physiology. *PloS Comput. Biol.* **8**, e1002571 (2012).

44. Succi, S. *The Lattice Boltzmann Equation for Fluid Dynamics and Beyond* (Clarendon Press, 2001).

45. Succi, S. *The Lattice Boltzmann Equation for Complex States of Flowing Matter.* (Oxford University Press, 2018).

46. Coveney, P., Díaz-Zuccarini, V., Hunter, P. & Viceconti, M. *Computational Biomedicine* (Oxford University Press, 2014).

47. Zacharoudiou, I., McCullough, J.W.S. & Coveney, P. V. Development and performance of HemeLB GPU code for human-scale blood flow simulation. *Comput. Phys. Commun.*, **282** 108548 (2023), https://doi.org/10.1016/j.cpc.2022.108548.

48. McCullough, J. W. S. et al. Towards blood flow in the virtual human: Efficient self-coupling of HemeLB. *Interface Focus* (December 2020). https://doi.org/10.1098/rsfs.2019.0119.

49. Hoekstra, A. G., Chopard, B., Coster, D., Zwart, S. P. & Coveney, P. V. Multiscale computing for science and engineering in the era of exascale performance. *Philos. Trans. R. Soc. A Math. Phys. Eng. Sci.* (2019). https://doi.org/10.1098/rsta.2018.0144.

50. McCullough, J.W.S. & Coveney, P. V. High fidelity physiological blood flow in patient-specific arteriovenous fistula for clinical applications. *Sci. Rep.* **11**(1), 22301 (2020).

51. Randles, A., Draeger, E. W., Oppelstrup, T., Krauss, L. & Gunnels, J. A. Massively parallel models of the human circulatory system. *Proc. Int. Conf. High Performance Comput., Networking, Storage and Analysis* (ACM, 2015). https://doi.org/10.1145/2807591.2807676.

52. Augustin, C. M. et al. A computationally efficient physiologically comprehensive 3D–0D closed-loop model of the heart and circulation. *Comput. Methods Appl. Mech. Eng.* **386**, 114092 (2021).

53. Holmes, R. Science fiction: The science that fed Frankenstein. *Nature* (2016). https://doi.org/10.1038/535490a.

Chapter 8: The Virtual Body

1. Sharma, D. et al. Technical note: In silico imaging tools from the VICTRE clinical trial. *Med. Phys.* (2019). https://doi.org/10.1002/mp.13674.

2. Hunter, P. J. & Borg, T. K. Integration from proteins to organs: The Physiome Project. *Nature Rev. Mol. Cell Biol.* (2003). https://doi.org/10.1038/nrm1054.

3. De Micheli, A. J. et al. Single-cell analysis of the muscle stem cell hierarchy identifies heterotypic communication signals involved in skeletal muscle regeneration. *Cell Rep.* (2020). https://doi.org/10.1016/j.celrep.2020.02.067.

4. Robinson, J. L. et al. An atlas of human metabolism. *Sci. Signal.* **13** (2020).

5. Liu, Q., Wu, B., Zeng, S. & Luo, Q. Human physiome based on the high-resolution dataset of human body structure. *Prog. Nat. Sci.* **18**, 921–925 (2008).

6. Benias, P. C. et al. Structure and distribution of an unrecognized interstitium in human tissues. *Sci. Rep.* (2018). https://doi.org/10.1038/s41598-018-23062-6.

7. Highfield, R. Cravings: How to trick the brain into thinking you're full. *Newsweek* (January 24, 2015). https://www.newsweek.com/2015/01/30/cravings-how-food-controls-our-brains-301495.html.

8. National Institutes of Health. Stimulating peripheral activity to relieve conditions (SPARC) (2022). https://commonfund.nih.gov/sparc/.

9. Costa, M. C. et al. Biomechanical assessment of vertebrae with lytic metastases with subject-specific finite element models. *J. Mech. Behav. Biomed. Mater.* (2019). https://doi.org/10.1016/j.jmbbm.2019.06.027.

10. Grünwald, A.T.D., Roy, S., Alves-Pinto, A. & Lampe, R. Assessment of adolescent idiopathic scoliosis from body scanner image by finite element simulations. *PloS One* **16**, e0243736 (2021).

11. Pitto, L. et al. SimCP: A simulation platform to predict gait performance following orthopedic intervention in children with cerebral palsy. *Front. Neurorobot.* **13**, 54 (2019).

12. Bozkurt, S. et al. Computational modelling of patient specific spring assisted lambdoid craniosynostosis correction. *Sci. Rep.* **10**, 18693 (2020).

13. Tawhai, M. H. & Hunter, P. J. Modeling water vapor and heat transfer in the normal and the intubated airways. *Ann. Biomed. Eng.* (2004). https://doi.org/10.1023/B:ABME.0000019180.03565.7e.

14. Lin, C. L., Tawhai, M. H., McLennan, G. & Hoffman, E. A. Characteristics of the turbulent laryngeal jet and its effect on airflow in the human intra-thoracic airways. *Respir. Physiol. Neurobiol.* (2007). https://doi.org/10.1016/j.resp.2007.02.006.

15. Burrowes, K. S., Swan, A. J., Warren, N. J. & Tawhai, M. H. Towards a virtual lung: Multiscale, multi-physics modelling of the pulmonary system. *Philos. Trans. R. Soc. A Math. Phys. Eng. Sci.* (2008). https://doi.org/10.1098/rsta.2008.0073.

16. Auckland Bioengineering Institute. MedTech CORE—lung simulation. https://sites.bioeng.auckland.ac.nz/medtech/lungs/.

17. Tawhai, M. H., Clark, A. R. & Chase, J. G. The Lung Physiome and virtual patient models: From morphometry to clinical translation. *Morphologie* (2019). https://doi.org/10.1016/j.morpho.2019.09.003.

18. Chan, H.-F., Collier, G. J., Parra-Robles, J. & Wild, J. M. Finite element simulations of hyperpolarized gas DWI in micro-CT meshes of acinar airways: Validating the cylinder and stretched exponential models of lung microstructural length scales. *Magn. Reson. Med.* **86**, 514–525 (2021).

19. Corley, R. A. et al. Comparative computational modeling of airflows and vapor dosimetry in the respiratory tracts of rat, monkey, and human. *Toxicol. Sci.* (2012). https://doi.org/10.1093/toxsci/kfs168.

20. Calmet, H. et al. Large-scale CFD simulations of the transitional and turbulent regime for the large human airways during rapid inhalation. *Comput. Biol. Med.* **69**, 166–180 (2016).

21. Franiatte, S., Clarke, R. & Ho, H. A computational model for hepatotoxicity by coupling drug transport and acetaminophen metabolism equations. *Int. J. Numer. Method. Biomed. Eng.* (2019). https://doi.org/10.1002/cnm.3234.

22. Muller, A., Clarke, R. & Ho, H. Fast blood-flow simulation for large arterial trees containing thousands of vessels. *Comput. Methods Biomech. Biomed. Engin.* (2017). https://doi.org/10.1080/10255842.2016.1207170.

23. Ho, H., Yu, H. B., Bartlett, A. & Hunter, P. An in silico pipeline for subject-specific hemodynamics analysis in liver surgery planning. *Comput. Methods Biomech. Biomed. Engin.* (2020). https://doi.org/10.1080/10255842.2019.1708335.

24. Ho, H., Bartlett, A. & Hunter, P. Virtual liver models in pre-surgical planning, intra-surgical navigation and prognosis analysis. *Drug Discovery Today: Disease Models* (2016). https://doi.org/10.1016/j.ddmod.2017.09.003.

25. Hoehme, S. et al. Prediction and validation of cell alignment along microvessels as order principle to restore tissue architecture in liver regeneration. *Proc. Natl. Acad. Sci. U. S. A.* (2010). https://doi.org/10.1073/pnas.0909374107.

26. Meyer, K. et al. A predictive 3D multi-scale model of biliary fluid dynamics in the liver lobule. *Cell Syst.* (2017). https://doi.org/10.1016/j.cels.2017.02.008.

27. Brown, S. A. et al. Six-month randomized, multicenter trial of closed-loop control in type 1 diabetes. *N. Engl. J. Med.* **381**, 1707–1717 (2019).

28. Sender, R., Fuchs, S. & Milo, R. Revised estimates for the number of human and bacteria cells in the body. *PloS Biol.* (2016). https://doi.org/10.1371/journal.pbio.1002533.

29. Amato, K. R. et al. Convergence of human and Old World monkey gut microbiomes demonstrates the importance of human ecology over phylogeny. *Genome Biol.* (2019). https://doi.org./10.1186/s13059-019-1807-z.

30. Mao, J. H. et al. Genetic and metabolic links between the murine microbiome and memory. *Microbiome* (2020). https://doi.org/10.1186/s40168-020-00817-w.

31. Muangkram, Y., Honda, M., Amano, A., Himeno, Y. & Noma, A. Exploring the role of fatigue-related metabolite activity during high-intensity exercise using a simplified whole-body mathematical model. *Informatics Med. Unlocked* **19**, 100355 (2020).

32. Brunk, E. et al. Recon3D enables a three-dimensional view of gene variation in human metabolism. *Nat. Biotechnol.* **36**, 272–281 (2018).

33. Thiele, I. et al. Personalized whole-body models integrate metabolism, physiology, and the gut microbiome. *Mol. Syst. Biol.* **16**, e8982 (2020).

34. Rzechorzek, N. M., et al. A daily temperature rhythm in the human brain predicts survival after brain injury. *Brain* (2022). https://doi.org/10.1093/brain/awab466.

35. Shapson-Coe, A. et al. A connectomic study of a petascale fragment of human cerebral cortex. *bioRxiv* (2021). https://doi.org/10.1101/2021.05.29.446289.

36. Abbott, A. How the world's biggest brain maps could transform neuroscience. *Nature* **598**, 22–25 (2021).

37. Arnulfo, G. et al. Long-range phase synchronization of high-frequency oscillations in human cortex. *Nat. Commun.* (2020). https://doi.org/10.1038/s41467-020-18975-8.

38. Wagstyl, K. et al. BigBrain 3D atlas of cortical layers: Cortical and laminar thickness gradients diverge in sensory and motor cortices. *PloS Biol.* (2020). https://doi.org/10.1371/journal.pbio.3000678.

39. Giacopelli, G., Migliore, M. & Tegolo, D. Graph-theoretical derivation of brain structural connectivity. *Appl. Math. Comput.* (2020). https://doi.org/10.1016/j.amc.2020.125150.

40. Wybo, W. A. et al. Data-driven reduction of dendritic morphologies with preserved dendro-somatic responses. *Elife* (2021). https://doi.org/10.7554/elife.60936.

41. Squair, J. W. et al. Neuroprosthetic baroreflex controls haemodynamics after spinal cord injury. *Nature* (2021). https://doi.org/10.1038/s41586-020-03180-w.

42. Neurotwin. https://www.neurotwin.eu/.

43. Highfield, R. How a magnet turned off my speech. *Daily Telegraph* (May 16, 2008). https://www.telegraph.co.uk/news/science/science-news/3342331/How-a-magnet-turned-off-my-speech.html.

44. Olmi, S., Petkoski, S., Guye, M., Bartolomei, F. & Jirsa, V. Controlling seizure propagation in large-scale brain networks. *PloS Comput. Biol.* (2019). https://doi.org/10.1371/journal.pcbi.1006805.

45. Aerts, H. et al. Modeling brain dynamics after tumor resection using The Virtual Brain. *Neuroimage* (2020). https://doi.org/10.1016/j.neuroimage.2020.116738.

46. Falcon, M. I. et al. Functional mechanisms of recovery after chronic stroke: Modeling with The Virtual Brain. *eNeuro* (2016). https://doi.org/10.1523/ENEURO.0158-15.2016.

47. Akil, H., Martone, M. E. & Van Essen, D. C. Challenges and opportunities in mining neuroscience data. *Science* (2011). https://doi.org/10.1126/science.1199305.

48. Pashkovski, S. L. et al. Structure and flexibility in cortical representations of odour space. *Nature* (2020). https://doi.org/10.1038/s41586-020-2451-1.

49. Turing, A. M. On computable numbers, with an application to the Entscheidungsproblem. *Proc. London Math. Soc.* (1937). https://doi.org/10.1112/plms/s2-42.1.230.

50. Church, A. An unsolvable problem of elementary number theory. *Am. J. Math.* (1936). https://doi.org/10.2307/2371045.

51. Penrose, R. What is reality? *New Sci.* (2006). https://doi.org/10.1016/S0262-4079(06)61094-4.

52. Hameroff, S. & Penrose, R. Consciousness in the universe: A review of the "Orch OR" theory. *Phys. Life Rev.* **11**, 39–78 (2014).

53. Krittian, S.B.S. et al. A finite-element approach to the direct computation of relative cardiovascular pressure from time-resolved MR velocity data. *Med. Image Anal.* (2012). https://doi.org/10.1016/j.media.2012.04.003.

Chapter 9: Virtual You 2.0

1. Babbage, C. *Passages from the Life of a Philosopher*. Cambridge Library Collection—Technology (Cambridge University Press, 2011). https://doi.org/10.1017/CBO9781139103671.

2. Kendon, V. M., Nemoto, K. & Munro, W. J. Quantum analogue computing. *Philos. Trans. R. Soc. A Math. Phys. Eng. Sci.* (2010). https://doi.org/10.1098/rsta.2010.0017.

3. Cowan, G.E.R., Melville, R. C. & Tsividis, Y.P.A VLSI analog computer/digital computer accelerator. *IEEE J. Solid-State Circuits* **41**, 42–53 (2006).

4. Pendry, J. B., Holden, A. J., Robbins, D. J. & Stewart, W. J. Magnetism from conductors and enhanced nonlinear phenomena. *IEEE Trans. Microw. Theory Tech.* **47**, 2075–2084 (1999).

5. Veselago, V. G. The electrodynamics of substances with simultaneously negative values of ε and μ. *Sov. Phys. Uspekhi* **10**, 509–514 (1968).

6. Pendry, J. B., Schurig, D. & Smith, D. R. Controlling electromagnetic fields. *Science* **312**, 1780–1782 (2006).

7. Molnár, B., Molnár, F., Varga, M., Toroczkai, Z. & Ercsey-Ravasz, M. A continuous-time MaxSAT solver with high analog performance. *Nat. Commun.* (2018). https://doi.org/10.1038/s41467-018-07327-2.

8. Alù, A. & Engheta, N. Achieving transparency with plasmonic and metamaterial coatings. *Phys. Rev. E—Stat. Nonlinear, Soft Matter Phys.* (2005). https://doi.org/10.1103/Phys RevE.72.016623.

9. Silva, A. et al. Performing mathematical operations with metamaterials. *Science* (2014). https://doi.org/10.1126/science.1242818.

10. Camacho, M., Edwards, B. & Engheta, N. A single inverse-designed photonic structure that performs parallel computing. *Nat. Commun.* **12**, 1466 (2021).

11. Estakhri, N. M., Edwards, B. & Engheta, N. Inverse-designed metastructures that solve equations. *Science* (2019). https://doi.org/10.1126/science.aaw2498.

12. Stroev, N. & Berloff, N. G. Discrete polynomial optimization with coherent networks of condensates and complex coupling switching. *Phys. Rev. Lett.* **126**, 50504 (2021).

13. Mead, C. Neuromorphic electronic systems. *Proc. IEEE* (1990). https://doi.org/10.1109 /5.58356.

14. Kumar, S., Williams, R. S. & Wang, Z. Third-order nanocircuit elements for neuromorphic engineering. *Nature* (2020). https://doi.org/10.1038/s41586-020-2735-5.

15. Chua, L. O. Memristor—the missing circuit element. *IEEE Trans. Circuit Theory* (1971). https://doi.org/10.1109/TCT.1971.1083337.

16. Strukov, D. B., Snider, G. S., Stewart, D. R. & Williams, R. S. The missing memristor found. *Nature* (2008) https://doi.org/10.1038/nature06932.

17. Joksas, D. et al. Committee machines—a universal method to deal with non-idealities in memristor-based neural networks. *Nat. Commun.* **11**, 4273 (2020).

18. Chua, L., Sbitnev, V. & Kim, H. Hodgkin-Huxley axon is made of memristors. *Int. J. Bifurcation and Chaos* (2012). https://doi.org/10.1142/S021812741230011X.

19. Brown, T. D., Kumar, S. & Williams, R. S. Physics-based compact modeling of electro-thermal memristors: Negative differential resistance, local activity, and non-local dynamical bifurcations. *Appl. Phys. Rev.* **9**, 11308 (2022).

20. Furber, S. *SpiNNaker: A Spiking Neural Network Architecture* (Now Publishers, 2020).

21. Yao, P. et al. Face classification using electronic synapses. *Nat. Commun.* (2017). https:// doi.org/10.1038/ncomms15199.

22. Pei, J. et al. Towards artificial general intelligence with hybrid Tianjic chip architecture. *Nature* (2019). https://doi.org/10.1038/s41586-019-1424-8.

23. Masanes, L., Galley, T. D. & Müller, M. P. The measurement postulates of quantum mechanics are operationally redundant. *Nat. Commun.* **10**, 1–6 (2019).

24. Rae, A. *Quantum Physics: Illusion or Reality?* 2nd ed. (Cambridge University Press, 2012).

25. Feynman, R. P. Simulating physics with computers. *Int. J. Theor. Phys.* (1982). https:// doi.org/10.1007/BF02650179.

26. Deutsch, D. Quantum theory, the Church-Turing principle and the universal quantum computer. *Proc. R. Soc. London, Ser. A Math. Phys. Sci.* (1985). https://doi.org/10.1098/ rspa.1985.0070.

27. Shor, P. W. Algorithms for quantum computation: Discrete logarithms and factoring. *Proc. 35th Ann. Sym. Found. Comput. Sci.*, 124–134 (1994). https://doi.org/10.1109/SFCS .1994.365700.

28. Lloyd, S. A potentially realizable quantum computer. *Science* (1993). https://doi.org /10.1126/science.261.5128.1569.

29. Salart, D., Baas, A., Branciard, C., Gisin, N. & Zbinden, H. Testing the speed of "spooky action at a distance." *Nature* (2008). https://doi.org/10.1038/nature07121.

30. Aspuru-Guzik, A., Dutoi, A. D., Love, P. J. & Head-Gordon, M. Simulated quantum computation of molecular energies. *Science* **309**, 1704–1707 (2005).

31. Ralli, A., Williams, M. I. & Coveney, P. V. A scalable approach to quantum simulation via projection-based embedding. *arXiv* (2022). https://doi.org/10.48550/arxiv.2203.01135.

32. Arute, F. et al. Quantum supremacy using a programmable superconducting processor. *Nature* (2019). https://doi.org/10.1038/s41586-019-1666-5.

33. Oak Ridge National Laboratory. Quantum supremacy milestone harnesses ORNL Summit supercomputer. Press release (October 23, 2019). https://www.ornl.gov/news /quantum-supremacy-milestone-harnesses-ornl-summit-supercomputer.

34. Oliver, W. D. Quantum computing takes flight: Expert insight into current research. *Nature* **574**, 487–488 (2019).

35. Pan, F., Chen, K. & Zhang, P. Solving the sampling problem of the Sycamore quantum circuits. https://arxiv.org/abs/2111.03011

36. Zhong, H.-S. et al. Quantum computational advantage using photons. *Science* (2021). https://doi.org/10.1126/science.abe8770.

37. Zhong, H.-S. et al. Phase-programmable Gaussian boson sampling using stimulated squeezed light. *Phys. Rev. Letters* **127**, 180502 (2021).

38. Banchi, L., Quesada, N. & Arrazola, J. M. Training Gaussian boson sampling distributions. *Phys. Rev. A* **102**, 12417 (2020).

39. Banchi, L., Fingerhuth, M., Babej, T., Ing, C. & Arrazola, J. M. Molecular docking with Gaussian boson sampling. *Sci. Adv.* **6**, eaax1950 (2020).

40. Jahangiri, S., Arrazola, J. M., Quesada, N. & Killoran, N. Point processes with Gaussian boson sampling. *Phys. Rev. E* **101**, 022134 (2020).

41. Peruzzo, A. et al. A variational eigenvalue solver on a photonic quantum processor. *Nat. Commun.* **5**, 4213 (2014).

42. Farhi, E., Goldstone, J. & Gutmann, S. A quantum approximate optimization algorithm. *arXiv* (2014). https://doi.org/10.48550/arxiv.1411.4028.

43. Ralli, A., Love, P., Tranter, A. & Coveney, P. Implementation of measurement reduction for the variational quantum eigensolver. *Phys. Rev. Res.* **3**, 033195 (2021).

Chapter 10: From Healthcasts to Posthuman Futures

1. Henshilwood, C. S. et al. An abstract drawing from the 73,000-year-old levels at Blombos Cave, South Africa. *Nature* (2018). https://doi.org/10.1038/s41586-018-0514-3.

2. Hublin, J. J. et al. New fossils from Jebel Irhoud, Morocco and the pan-African origin of *Homo sapiens*. *Nature* (2017). https://doi.org/10.1038/nature22336.

3. Marchant, J. *The Human Cosmos: A Secret History of the Stars* (Canongate, 2020).

4. Aubert, M. et al. Pleistocene cave art from Sulawesi, Indonesia. *Nature* (2014). https:// doi.org/10.1038/nature13422.

5. Conard, N. J. A female figurine from the basal Aurignacian of Hohle Fels Cave in southwestern Germany. *Nature* (2009). https://doi.org/10.1038/nature07995.

6. McCoid, C. H. & McDermott, L. D. Toward decolonizing gender: Female vision in the Upper Paleolithic. *Am. Anthropol.* (1996). https://doi.org/10.1525/aa.1996.98.2.02a00080.

7. Highfield, R. *The Physics of Christmas: From the Aerodynamics of Reindeer to the Thermodynamics of Turkey* (Little, Brown, 1998).

8. Longest, P. W. et al. Use of computational fluid dynamics deposition modeling in respiratory drug delivery. *Expert Opin. Drug Deliv.* **16**, 7–26 (2019).

9. Wouters, O. J., McKee, M. & Luyten, J. Estimated research and development investment needed to bring a new medicine to market, 2009–2018. *J. Am. Med. Assoc.* (2020). https:// doi.org/10.1001/jama.2020.1166.

10. Turner, E. H., Matthews, A. M., Linardatos, E., Tell, R. A. & Rosenthal, R. Selective publication of antidepressant trials and its influence on apparent efficacy. *N. Engl. J. Med.* (2008). https://doi.org/10.1056/nejmsa065779.

11. Jefferson, T. et al. Neuraminidase inhibitors for preventing and treating influenza in adults and children. *Cochrane Database Sys. Rev.* (2014). https://doi.org/10.1002/14651858.CD008965.pub4.

12. Kam-Hansen, S. et al. Altered placebo and drug labeling changes the outcome of episodic migraine attacks. *Sci. Transl. Med.* (2014). https://doi.org/10.1126/scitranslmed.3006175.

13. Leucht, S., Helfer, B., Gartlehner, G. & Davis, J. M. How effective are common medications?: A perspective based on meta-analyses of major drugs. *BMC Med.* (2015). https://doi.org/10.1186/s12916-015-0494-1.

14. Boyle, E. A., Li, Y. I. & Pritchard, J. K. An expanded view of complex traits: From polygenic to omnigenic. *Cell* **169**, 1177–1186 (2017).

15. MacKenzie, A. & Kolb, A. The human genome at 20: How biology's most-hyped breakthrough led to anticlimax and arrests. *Phys.org* (2021). https://phys.org/news/2021-02-human-genome-biology-most-hyped-breakthrough.amp.

16. ENCODE Project Consortium. The ENCODE (encyclopedia of DNA elements) Project. *Science* **306**, 636–640 (2004).

17. Hall, W. D., Mathews, R. & Morley, K. I. Being more realistic about the public health impact of genomic medicine. *PloS Med.* (2010). https://doi.org/10.1371/journal.pmed.1000347.

18. McGuire, A. L. et al. The road ahead in genetics and genomics. *Nature Rev. Genetics* (2020). https://doi.org/10.1038/s41576-020-0272-6.

19. Bhati, A. P. et al. Pandemic drugs at pandemic speed: Infrastructure for accelerating COVID-19 drug discovery with hybrid machine learning- and physics-based simulations on high-performance computers. *Interface Focus* **11**, 20210018 (2021).

20. Wan, S., Bhati, A. P., Wade, A. D., Alfè, D. & Coveney, P. V. Thermodynamic and structural insights into the repurposing of drugs that bind to SARS-CoV-2 main protease. *Mol. Sys. Des. Eng.* (2021). https://doi.org/10.33774/chemrxiv-2021-03NRL-V2.

21. Price, N. D. et al. A wellness study of 108 individuals using personal, dense, dynamic data clouds. *Nat. Biotechnol.* **35**, 747–756 (2017).

22. Earls, J. C. et al. Multi-omic biological age estimation and its correlation with wellness and disease phenotypes: A longitudinal study of 3,558 individuals. *J. Gerontol. Ser. A* **74**, S52–S60 (2019).

23. Niederer, S. A., Sacks, M. S., Girolami, M. & Willcox, K. Scaling digital twins from the artisanal to the industrial. *Nat. Comput. Sci.* **1**, 313–320 (2021).

24. Voigt, I. et al. Digital twins for multiple sclerosis. *Front. Immunol.* **12**, (2021).

25. Royal Society. Data management and use: Governance in the 21st century. Joint report by the British Academy and the Royal Society (June 2017). https://royalsociety.org/topics-policy/projects/data-governance/.

26. Nordling, L. A fairer way forward for AI in health care. *Nature* **573**, S103–S105 (2019).

27. UN OHCHR. Artificial intelligence risks to privacy demand urgent action—Bachelet. Press release (September 15, 2021). https://www.ohchr.org/EN/NewsEvents/Pages/media.aspx?IsMediaPage=true.

28. de Kerckhove, D. The personal digital twin, ethical considerations. *Philos. Trans. R. Soc. A Math. Phys. Eng. Sci.* **379**, 20200367 (2021).

29. Bruynseels, K., de Sio, F. S. & van den Hoven, J. Digital twins in health care: Ethical implications of an emerging engineering paradigm. *Front. Genet.* (2018). https://doi.org/10.3389/fgene.2018.00031.

Appendix: Towards a Virtual Cosmos

1. Shirasaki, M., Sugiyama, N. S., Takahashi, R. & Kitaura, F.-S. Constraining primordial non-Gaussianity with postreconstructed galaxy bispectrum in redshift space. *Phys. Rev. D* **103**, 23506 (2021).

2. Coveney, P. & Highfield, R. *Frontiers of Complexity: The Search for Order in a Chaotic World* (Faber, 1996).

3. Hardy, J., de Pazzis, O. & Pomeau, Y. Molecular dynamics of a classical lattice gas: Transport properties and time correlation functions. *Phys. Rev. A* **13**, 1949–1961 (1976).

4. Wolfram, S. Cellular automaton fluids 1: Basic theory. *J. Stat. Phys.* **45**, 471–526 (1986).

5. Liberati, S. Tests of Lorentz invariance: A 2013 update. *Class. Quantum Gravity* **30**, 133001 (2013).

6. Chou, A. S. et al. First measurements of high frequency cross-spectra from a pair of large Michelson interferometers. *Phys. Rev. Lett.* **117**, 111102 (2016).

7. Wolfram, S. *A Project to Find the Fundamental Theory of Physics* (Wolfram Media, 2020).

8. Gorard, J. Some relativistic and gravitational properties of the Wolfram model. *Complex Syst.* **29**, 599–654 (2020).

9. Gorard, J., Namuduri, M. & Arsiwalla, X. D. *ZX-Calculus and Extended Hypergraph Rewriting Systems I: A Multiway Approach to Categorical Quantum Information Theory. arXiv* (2020). https://doi.org/10.48550/arxiv.2103.15820.

Index

A page number in *italics* refers to a figure.